I0091799

Manly Miles

Stock-Breeding

A practical Treatise

Manly Miles

Stock-Breeding
A practical Treatise

ISBN/EAN: 9783337143596

Printed in Europe, USA, Canada, Australia, Japan

Cover: Foto ©berggeist007 / pixelio.de

More available books at **www.hansebooks.com**

STOCK-BREEDING:

A PRACTICAL TREATISE

ON THE APPLICATIONS OF THE

LAWS OF DEVELOPMENT AND HEREDITY TO THE IMPROVEMENT AND BREEDING OF DOMESTIC ANIMALS.

BY

MANLY MILES, M. D.,

LATE PROFESSOR OF AGRICULTURE IN THE MICHIGAN STATE AGRICULTURAL COLLEGE.

NEW YORK:

D. APPLETON AND COMPANY,

549 AND 551 BROADWAY.

1879.

PREFACE.

It is somewhat remarkable, in this book-making age, that there is no systematic work accessible to the student in which the known facts and principles of the art of improving and breeding domestic animals are presented, in convenient form, for study and reference, notwithstanding the importance of live-stock to the farmer, and the wonderful progress that has been made in its improvement since the time of Bakewell.

The present attempt to supply this want has been made in response to the repeated solicitations of persons interested in stock-breeding, who have attended my lectures on this subject, in various places, for several years past.

In a popular exposition of the principles of an art that is almost exclusively based upon the experience of practical men there is little opportunity for originality, aside from the classification and arrangement of facts, and the inferences, in some instances, that may be drawn from them in explaining the practice of the most successful breeders.

It is believed that a systematic statement of what is already known in the practice of the art is of greater importance, at the present time, than any new truths, as it must furnish the only consistent foundation for future progress and improvement.

The numerous cases that have been collected to illustrate the various topics under discussion have been compiled, as far as possible, from original sources and presented in their original form—references, in nearly all cases, being given to the works from which they are quoted.

This feature of the work will be of interest to the student who wishes to study the subject in greater detail, as it will, to some extent, serve as an index to authorities that may be profitably consulted.

In the limits of a popular work it is of course impossible to treat each topic exhaustively, and the attempt has been made to present only such an outline of the principles of the art as would be required in a text-book for students, or a work of reference for farmers.

The acknowledgments of the author are due to the well-known animal-artist John R. Page, of Sennett, New York, for the spirited illustrations in the chapter on " Form," all of which are from life, with the exception of Fig. 8, which is after a sketch by the Hon. Francis Rotch.

It is to be hoped that the *résumé* of cases here presented may lead breeders to recognize the importance of placing on record the additional facts, from their own experience, that are required for a more complete discussion of the subjects treated in this volume.

LANSING, MICHIGAN, *July* 20, 1878.

CONTENTS.

PRINCIPLES OF STOCK-BREEDING.

CHAPTER I.

BREEDING AS AN ART.

THE art of breeding domesticated animals, for the various purposes to which they are adapted, has been practised from the earliest times.

The oldest writers on agriculture gave directions for the breeding and improvement of cattle, and some of their maxims are often repeated by modern authorities as the best practical guides to the farmer.

It has long been known that the characteristics of parents were transmitted to their offspring, and the results of observation were tersely expressed in the familiar aphorism, "like produces like." As a natural corollary of this generally-accepted law of the animal organization, the rule "breed from the best" very early found a place among the approved maxims of the art.

The principles of breeding, up to the time of Bakewell, were essentially comprised in these two apothegms; but it is evident, from the practice of breeders, that they did not fully appreciate the extended applications of these empirical expressions, that represent the fundamental principles of the best modern practice.

The early breeders, like many at the present time, had no consistent system of selection. The "best" of any given selection for breeding, made in accordance with the time-honored rule, differed in all essential details of form and quality from the "best" that were selected at another time.

Their standard of excellence was, in fact, constantly changing, so that no real progress in the development of the most valuable qualities could be made.

Shortly after the middle of the last century, Robert Bakewell, of Dishley Grange, Leicestershire, England, originated a new system which he successfully practised in the improvement of Leicester sheep, Long-horn cattle, and Black cart-horses. His belief that the familiar maxim, "like begets like," was not limited to a general similarity of the offspring to the parent, but extended to the minutest details of the organization, led him to adopt for his guidance a definite standard of excellence representing the form and internal qualities that were best adapted to the highest development of the animal for a special purpose.

His critical study of the form and proportions of animals, and their relations to the most desirable qualities, enabled him to develop an ideal model of perfection, that he kept constantly in view when making his selections for breeding.

In his sheep and cattle he endeavored to secure a large proportion of choice parts in the carcass, a superior quality of flesh, with a tendency to early maturity, and uniformity in the transmission of their most valuable qualities to their offspring.

Beauty in the form and proportions of his animals was always made to contribute to the development of useful characters. Mr. Bakewell's success in the improvement of the animals he was breeding must be attributed to the exercise of a combination of talents that would have made him eminent in any profession or pursuit.

A correct and well-trained eye enabled him to detect the slightest variations of form; and these, from his knowledge of the animal organization, obtained by long-continued and systematic observation, he associated with the correlated qualities they represented.

Relying upon his own good judgment, which was not biased by non-essential conditions or fanciful theories, he not only accepted all that was consistent in the received rules of the art, but established new principles of the greatest practical importance.

He seems to have been apt in tracing the relations of cause and effect, and methodical and persevering in the execution of his well-considered plans for improvement.

With the spirit of a true artist, he endeavored to mould the plastic forms of his animals to give expression to his ideal conception of the qualities that constitute perfection.

The method of Bakewell has been successfully practised by other able men; and we now have, as the result of their labors, a variety of improved breeds of remarkable excellence, each differing from the others in the particular characters that adapt them to special conditions and purposes.

Success in the breeding of live-stock, as in all other departments of farm management, must be measured by the actual value of the products, and the profits that may be derived from them.

The relative value of animals depends upon their adaptation to a particular purpose, and the returns they make for feed consumed. It is evident that, where a particular form of animal product is the leading object, the greater value will be placed upon the animal that excels in its production. Excellence in other directions may be desirable, but it will not compensate for a deficiency in the special qualities required.

The return obtained, in any form of animal product, for feed consumed, is of the first importance in estimating the value of animals. Mr. Bakewell regarded live-stock as machines for converting the vegetable products of the farm into animal products of greater value; and Sir John Sinclair expresses the same idea when he says, "Under the head of live-stock are comprehended the various sorts of domesticated animals which are employed by man as instruments for converting to his use, either by labor or otherwise, those productions of the soil which are not immediately applicable to supply his wants in their natural state." [1]

The animal that furnishes the largest amount and the best quality of the desired animal product, from a given amount of food, would undoubtedly be the

[1] "Code of Agriculture," p. 84.

"best;" or, looking more particularly at the activity of the animal machinery, it might be said that the animal that converts the largest amount of food into animal products of the best quality, with the least possible waste of material, would be the most valuable.

It is often assumed that animals that eat but little are the most profitable, but this error is evidently founded on mistaken notions of the functions of animal life, and the true place that they occupy in the economy of the farm. A machine that will convert the largest amount of raw material into the desired product, with the least possible wear, and the least expenditure of fuel to furnish the required motive-power, would be more valuable than one that required less fuel, but in which the capacity for efficient work was diminished in a greater ratio.

The repairs of the animal machine are made at the expense of food consumed, and, if the animal is capable of digesting and assimilating only what is required for this purpose, it would be comparatively worthless, as a profit can only be obtained from the food assimilated in excess of this amount.

In my experiments in feeding swine, the best returns for feed consumed were obtained when the animals ate the most in proportion to their live weight, and this is undoubtedly the rule in stock-feeding. This is readily explained by the fact that, when a large amount of food is consumed by an animal, providing it is capable of digesting and assimilating it, the *proportion* of food required to supply the waste of the tissues and keep the animal machinery in working order is less than when a smaller amount is consumed.[1]

It will not, then, be desirable to breed animals that eat but little, as we cannot reasonably expect them to give as large a proportionate return for feed consumed as those that have efficient digestive organs of greater capacity.

A comparison of the results obtained with different animals is generally neglected by farmers, and they therefore make too little difference in the price of their best animals, that are capable of returning a fair profit on the food consumed, and those of inferior quality, that do not, perhaps, pay for their keep.

The great difference in the relative value of animals will be best shown by a few illustrations.

One cow yields five pounds of butter a week, which, at twenty cents a pound, would be one dollar, or twenty dollars for twenty weeks, which we will assume is the period of usefulness for the year.

Another yields eight pounds of butter a week, worth one dollar and sixty cents, and the total return for twenty weeks would be thirty-two dollars, or twelve dollars more than was realized from the first cow. This difference represents the interest on one hundred and twenty dollars, at ten per cent. for the year. On the same basis, a cow yielding ten pounds of butter a week would earn forty dollars in twenty weeks, or twenty dollars more than the first cow, and this difference would be the interest on two hundred dollars for the year.

Three pounds of wool from one sheep, at fifty

[1] "Michigan Agricultural Report," 1873, p. 120.

cents per pound, would bring one dollar and fifty cents; and six pounds from another, at the same price, would bring three dollars, a difference of one dollar and fifty cents, or the interest on fifteen dollars. Even if it is claimed that the animals giving the greatest return consume considerably more food than the inferior ones, there would still remain a great difference in the profits of their products.

If in the same way we estimate the relative value of sires, by comparing the qualities of their offspring, it would be seen that one capable of increasing the value of the flock or herd would be well worth a good price; while another, that could not be relied on to impress any good qualities upon his offspring, would be dear at any price.

The object of the art of breeding is the improvement of animals in those qualities that have a definite value, among which are the production of meat, and milk, and wool, and labor.

Breeders who have been the most successful in improving the various pure breeds have endeavored to obtain the highest development of some one of these qualities; while the others, which they looked upon as of secondary importance, have been quite generally neglected.

It must not, however, be assumed that these qualities are absolutely incompatible, so that a high degree of excellence in two or more of them cannot be obtained in the same animal; but it is undoubtedly easier to secure an extraordinary development of a single character than to obtain the same degree of excellence in two or more at the same time.

When the entire energies of the system are acting in a particular direction, as they must do to insure the highest development of a single quality, there is no residuum of force for the development of other qualities that are not strictly correlated with the one that is made dominant.

The modern art of breeding is founded on the practice of the most successful breeders, and its rules have been almost exclusively empirical in their origin.

The science of physiology, by explaining the principles on which many of these rules are based, has defined the limits of their applications with greater exactness, and suggested new fields for investigation.

In its progress the art has, however, kept constantly in advance of the allied science of physiology, which it has aided in developing by presenting, in its definite facts, the required data for successful scientific study.[1]

With the progress of knowledge, the unexplained precepts of the art are gradually diminishing, and the many theories that have been framed from partial views of the truth must be replaced by consistent principles of general application.

The inferior quality of the live-stock on the farms throughout the country shows that the relations of the art of breeding to the practice of agriculture have been too generally overlooked by farmers.

Looking upon live-stock as a special interest, and

---

[1] Whewell remarks that "in all cases the arts are prior to the related sciences. Art is the parent, not the progeny, of science: the realization of principles in practice forms part of the prelude, as well as the sequel, of theoretical discovery" ("History of the Inductive Sciences," vol. i., p. 240).

referring to their past experience with animals that were entirely unfitted for any useful purpose, it is not strange that the assumption that "live-stock will not pay" is so often repeated.

The same opinion seemed to prevail among the farmers of Rome in the first century, and Columella pointed out to them the fallacy of this prejudice against one of the most important interests of the farm.[1]

Conrad Heresbach, quoting *Fundanius* in Varro, compares the tillage of the soil and the interest in live-stock to two instruments in an orchestra, each differing in sound; and he terms "the grazier's trade the treble, and the tiller's occupation the base," each aiding in the harmony as a whole.[2]

Fitzherbert expresses the same idea when he says, " An husbande cannot well thryve by his corne without he have other cattell, nor by his cattell without corne, for els he shall be a byer, a borrower, or a beggar."[3]

George Culley, in his valuable treatise on live-stock, says, "According to the present improved system of farming there is such a connection between the cultivation of the ground and breeding, rearing, and fattening cattle, sheep, and other domestic animals, that a man will make but an indifferent figure in rural affairs if he does not understand the latter as well as the former."[4]

[1] "Columella of Husbandry," book vi., p. 255.
[2] "Foure Bookes of Husbandrie" (1586), p. 111.
[3] "Boke of Husbandry" (1532), p. 31 (reprint, 1767).
[4] "Live-Stock," by Culley, fourth edition, 1807, p. 1.

The statements of these writers were not made on theoretical grounds, but represented the opinions of those who were most successful in their system of practice.

At the present time farmers throughout the country should give greater attention to live-stock, as the first step in the improvement of their system of farming.

Animals of the best quality, that are adapted to the conditions of the farm and the particular purpose that the system of management demands, will yield profitable returns for the feed consumed, and furnish the best means of enriching the soil for the growing of grain.

The principles that guide the breeder of pure-bred stock are likewise applicable to the improvement of the common stock of the farm.

It is not to be expected that all persons will be equally successful in producing animals of extraordinary merit, but it is, nevertheless, true that a careful study of the principles of the art, which are easily understood, will enable the farmer to make improvements in his stock that will add largely to his profits.

CHAPTER II.

THE inheritance by the offspring of the characters of the parents, at the time of procreation, has been generally accepted as a law of the animal organization.

Although there are many apparent exceptions to this law, an examination of all the facts relating to the hereditary transmission of structure and qualities will, however, show that it is not only constant in its action, but extends to every feature of the organization, and that the supposed exceptions are the result of the predominant influence of other laws that obscure the hereditary tendency, for the time being, without wholly suppressing it.

The resemblance of offspring to parents, so frequently remarked, is not, as might at first be supposed, confined to the external and more obvious characters, but manifests itself in the internal structure and functional activity of the system. In fact, at the moment of birth, the sum of the characters and qualities of the young animal have been derived from its parents, and we shall find reason for the belief that they include every peculiarity in the organization of both parents.

The nervous system and mental condition, the organs of nutrition and reproduction, the habits, predispositions, and temperament, the bones, the muscles, and the powers of endurance, that characterize the parents, are all reproduced in the offspring without essential change in their characteristics.

Illustrations, drawn from the different departments of organic life, will serve to show the extent and persistent action of the law of heredity, and aid us in determining its applications in the breeding of domestic animals.

In the geological formations, representing immense periods of time, fossil species and generic forms present the same essential characters throughout their entire range.

The various species of wild animals are readily recognized wherever found, and the lapse of time represented in the historic period has made no appreciable change in their characters. The animals that have been preserved in the monuments of Egypt for thousands of years are essentially the same as those now found on the borders of the Nile.[1]

So far as the art of breeding is concerned, a consideration of the various theories of evolution can be of no practical value, and the observed repetition of generic and specific forms may be assumed to represent a constancy in the inherited characteristics of animals.

The cycle of changes through which the embryo passes in the process of development remains the

---

[1] Colin, "Physiologie Comparée," tome ii., p. 533; Darwin's "Animals and Plants under Domestication," vol. i., pp. 30–60.

same in the various species, while the disappearance
of organs that serve a temporary purpose takes place
at the same period of growth.   The races of men,
when pure, are readily distinguished by peculiarities
in complexion, features, and general organization.

The Jews and the gypsies have been cited as illustrations of the hereditary transmission of the peculiarities of a race, as they do not intermarry with other
families, and their distinguishing characteristics have
remained the same for centuries.[1]

The uniformity observed in the various breeds of
domestic animals is the result of the inheritance of
the characters that adapt them to the conditions under
which they have originated.   In the improved breeds
advantage is taken of the hereditary transmission of
certain family peculiarities that have been ingrafted
upon those of the original breed.   As the origin and
development of these improved characters have no
relation to our present subject, they will be considered
in another chapter.

In almost every breed there are favorite families,
that are prized by breeders for the persistence with
which they stamp their peculiar characters upon their
offspring.

The breeders of sheep will call to mind the influence of the Ellman and the Webb sorts in the improvement of the Southdowns, and of the Dishley
family in the development of the Leicesters.

The different *cabañas* of merino sheep in Spain,

---

[1] "Heredity," by Ribot, pp. 112–114 ; "Journal of the Royal Agricultural Society," vol. xiv., p. 106 ; Goodale's " Principles of Breeding,"
p. 23 ; Anderson's " Recreations in Agriculture," vol. i., p. 71.

that were kept distinct for many years, were characterized by peculiarities that were uniformly inherited.[1]

Of the offshoots of the original Spanish-merino
sheep, the Silesian family bred in Europe, and the
Rich and the Hammond families in the United States,
furnish further illustrations of the hereditary transmission of family characters.

Of cattle, the Booth and the Bates families of
Short-horns, the Quartley family of Devons, the Dishley family of Long-horns, and the Ben Tompkins sort
of the Herefords, may be mentioned as among the
favorites of breeders, on account of the marked heredity of their peculiar qualities.

Muscular strength, in connection with remarkable
powers of endurance, is frequently observed in particular families. "In ancient times there were families of athletes, and there have been families of prize-
fighters. The recent researches of Galton as to wrest-
lers and oarsmen show that the victors generally
belong to a small number of families, among whom
strength and skill are hereditary."[2]

The large proportion of successful racers tracing
their ancestry to Herod and Eclipse, not in a single
line only, but in several, furnish a good illustration of
the hereditary transmission of muscular power. It is
said that "Eclipse begot 334 and Herod 497 winners."[3]

[1] Livingstone on "The Sheep," p. 21; Morrell's "American Shepherd," pp. 71-75; Randall's "Practical Shepherd," p. 14; Youatt on
"Sheep," p. 156.

[2] "Heredity," by Ribot, p. 6.

[3] "The Horse," by Stonehenge, American edition, p. 142; Stone-
henge, "British Rural Sports," p. 282; "The Horse of America," by
Frank Forrester, vol. ii., p. 265; Darwin's "Animals and Plants under
Domestication," vol. ii., p. 21.

The most successful American trotting-horses are said to belong to but three families, and of these the Messenger is thought to be the best, as it has furnished a larger number of fast trotters than any other.[1]

There are families that inherit that peculiar organic structure of ear, nervous system, and vocal organs, that gives rise to what is recognized as musical talent. One of the most remarkable instances of this form of heredity on record is that of the family of Sebastian Bach. "It began in 1550, and continued through eight generations. . . . During a period of nearly two hundred years this family produced a multitude of artists of the first rank. . . . Its head was Weit Bach, a baker of Presburg, who used to seek relaxation from labor in music and song. He had two sons, who commenced that unbroken line of musicians of the same name that for nearly two centuries overran Thuringia, Saxony, and Franconia. . . . They were all organists or church-singers. . . . In this family are reckoned twenty-nine eminent musicians."[2]

The feeding quality, or tendency to lay on fat, which is one of the most important characteristics of the meat-producing breeds of animals, is also hereditary. In each distinctive breed, where the production of meat is the leading quality, there are certain families that excel in this direction.

It has been claimed that the predisposition to obesity is so strong in many cases that it is observed

[1] "Horse Portraiture," by Simpson, p. 303.

[2] Ribot on "Heredity," p. 63 ; Carpenter's "Mental Physiology," p. 273.

even under the disadvantages of privation and hard labor.[1]

The duration of the life of an individual is determined, to a great extent at least, by inheritance. The members of some families die at an early age, while in other families a ripe old age may be reasonably expected. The life-tables that have been constructed show the average expectation of life of the masses; while the expectation of life of an individual can only be approximately determined by the age attained by his ancestors.

Darwin mentions "the case of four brothers who died between the ages of sixty and seventy, in the same highly-peculiar comatose state.[2] . . .

"It is now generally understood that longevity depends far less on race, climate, profession, mode of life, or food, than on hereditary transmission."

There are long-lived families under what would be considered unfavorable conditions for longevity, while other families are short-lived under the most favorable conditions for the promotion of health. "The average of life," says Dr. Lucas, "plainly depends on locality, hygiene, and civilization, but individual longevity is entirely exempt from these conditions. Everything tends to show that long life is the result of an internal principle of vitality, which privileged individuals receive at their birth. It is so deeply imprinted in their nature as to make itself apparent in every part of the organization."[3]

[1] "Heredity," by Ribot, p. 3; Colin, "Physiologie Comparée," tome ii., p. 534.

[2] "Animals and Plants under Domestication," vol. ii., p. 28.

[3] "Heredity," by Ribot, p. 5; Smith's "Physical Indications of Longevity," p. 3.

The remarkable difference that is observed in the fecundity, not only of individuals but of classes and families, is undoubtedly owing to inherited peculiarities of the system. As a rule, the lower groups of animals present a greater activity of the reproductive powers than the higher. Among the vertebrates the oviparous classes are more prolific than the viviparous. Certain families are noted for their fecundity, while in others it is rare to find an individual that has many descendants.

Girou relates the case of a mother who gave birth to twenty-four children, among them five girls, who in turn gave birth to forty-six children in all. The daughter of this woman's son, while still young, gave birth to her sixteenth child.

In some families inherited fecundity has been observed for five or six generations. "The sons, daughters, and grandchildren, of a couple who were the parents of nineteen children, were nearly all gifted," says Lucas, "with the same fecundity." [1]

Those familiar with the various breeds of domestic animals will call to mind many cases that illustrate the heredity of the procreative powers.

Of the high-bred families of the improved breeds, some are remarkably prolific, while others are almost uniformly deficient in this important quality.

The imported Short-horn cow, Young Mary, by Jupiter (2170), had fourteen heifer-calves and one bull, and died at the age of twenty-one years. Her offspring were almost without exception re-

---

[1] "Heredity," by Ribot, p. 4 ; "Cyclopædia of Anatomy and Physiology," vol. ii., p. 471.

markably prolific. Mr. Lewis F. Allen, the editor of
the "American Short-Horn Herd-Book," says, "More
herd-book pedigrees run to Young Mary than any
other half-dozen cows on record." [1]

It is generally admitted by physiologists that the
mental peculiarities of an individual are determined,
to a great extent, by hereditary influences.

Dr. Carpenter says : " The view of the relation of
mental habits to peculiarities of bodily organization,
whether *congenital* or *acquired*, must be extended to
that remarkable hereditary transmission of psychical
character which presents itself under circumstances
that entirely forbid our attributing it to any agency
that can operate subsequently to birth, and which it
would seem impossible to account for on any other
hypothesis than that the 'formative capacity' of the
germ, in great degree, determines the subsequent development of the brain, as of other parts of the body,
and (through this) its mode of activity. . . . And this
formative capacity, which is the physiological expression of what is commonly spoken of as the 'original
constitution' of each individual, is essentially determined by the conditions, dynamical and material, of
the parent organisms." [2]

In domestic animals it is a matter of common observation that the temper or disposition, and other
mental peculiarities of individuals, are determined

---

[1] " History of Short-Horns," p. 217.   For further illustrations, *see*
the chapter on " Fecundity."

[2] " Mental Physiology," pp. 367, 368 ; " Cyclopædia of Anatomy and
Physiology," vol. ii., p. 471 ; Carpenter's " Human Physiology," p. 817.
Ribot, in his work on " Heredity," gives an extended discussion of the
hereditary transmission of mental peculiarities.

by inheritance. The peculiar imperfection of vision manifested in the inability to distinguish colors— popularly known as color-blindness—is also hereditary. Of the many instances on record of the hereditary transmission of this defect, the following is perhaps the most remarkable. Dr. Pliny Earle says: "My maternal grandfather and two of his brothers were characterized by it, and among the descendants of the first-mentioned there are *seventeen* persons in whom it is found. I have not been able to extend my inquiries among the collateral branches of the family, but have heard of one individual, a female, in one of them, who was similarly affected. . . . Nothing is known of the first generation (of five) in regard to the power of the perception of colors. In the second, of a family consisting of seven brothers and eight sisters, three of the brothers—one of whom, as before mentioned, was the grandfather of the writer—had the defect in question. In the third generation, consisting of the children of the grandfather aforesaid, of three brothers and four sisters, there was no one whose ability to distinguish colors was imperfect. In the fourth generation, the first family includes five brothers and four sisters, of whom two of the former have the defect. In the second family there was but one child, whose vision was normal. In the third there were seven brothers, of whom four had the defect. In the fifth, seven sisters and three brothers, of all of whom the vision is perfect in regard to color. In the sixth, four brothers and five sisters, of whom two of each sex have the defect. In the seventh, two brothers and three sisters—both of the former have the defect. In the eighth there was no issue, and in the ninth there are two sisters, both of them capable of appreciating colors. Of the fifth generation, the defective perception has hitherto been detected in but two of the families. In one of them, consisting of three brothers and three sisters, one of the brothers has the defect, and in the other, a male, an only child, is similarly affected." [1]

The peculiar condition of the lens of the eye observed in short-sighted people is hereditary, as is also the opposite defect, giving what is known as long sight. [2]

"Day-blindness, or imperfect vision under a bright light, is inherited, as is night-blindness, or an incapacity to see except under a strong light. A case has been recorded by M. Currier of this latter defect having affected eighty-five members of the same family during six generations." [3]

Dr. Earle likewise gives the case of a family in which a defective musical ear is associated with an imperfect appreciation of colors.

In a family of my acquaintance a peculiarity in the walk is hereditary.

Ribot says that in some families the hair turns gray in early youth, and similar cases have come under my own observation.

The loss of the teeth, when a particular age is

[1] *American Journal of the Medical Sciences*, vol. xxxv., p. 317; quoted in "Cyclopædia of Anatomy and Physiology," vol. iv., p. 1453.

[2] Darwin, "Animals and Plants under Domestication," vol. ii., p. 17.

[3] Darwin, *loc. cit.*, vol. ii., pp. 19, 269.

reached, is also an hereditary character in many families.

Without a further enumeration of details, it may be said that every peculiarity of the animal organization is influenced by heredity.

CHAPTER III.

ANY abnormal peculiarities of the animal organization, constituting disease, whether of structure or function, are liable to be transmitted from parent to offspring.

When a disease is characterized by obvious structural changes in any part of the system, its heredity is seldom called in question; but when it consists in a simple derangement of function, without any apparent indications of structural transformation, its hereditary character is frequently overlooked. As the progress of physiological science, however, makes us better acquainted with the minute structure of the various organs of the body, and the relations of such structure to their activity, the cases of functional disturbance that are not known to be accompanied by corresponding changes in structure have rapidly diminished, and that to so great an extent that it seems probable that all indications of disease are the result of some structural modification of the organs involved.

The hereditary transmission of some peculiarity in the performance of the function of an organ, without apparent structural change, is perhaps not more difficult to understand than the heredity of habits that, in themselves, are not beneficial. It may, in fact, be said that any peculiarity in the functional activity of an organ, if long continued, may result in a habit of the system which the offspring will in all probability inherit.

Hereditary disease may make its appearance at the time of birth, when it is said to be congenital, or a considerable length of time may elapse before any indications of its presence are observed. In the latter case a *predisposition* or tendency to the disease is said to be inherited, which often requires some external exciting cause for its full development.[1]

There are certain diseases that are transmitted with greater uniformity than others; yet a predisposition to almost every known form of disease is likely to become hereditary, even if the influence that determines its transmission is not sufficiently intense to render it congenital. It is not my purpose to describe or even enumerate all the diseases that are known to be hereditary, but to notice only those that illustrate the laws of hereditary transmission, or that, from their frequent occurrence, are of particular interest to breeders of domestic animals. Under the general term "scrofula," a great variety of disorders are included, all of which are characterized by a perversion of the nutritive functions, and the formation of peculiar tumors, called tubercles, in the various organs of the body. The most common forms of scrofulous

<hr>

[1] Adams on "Hereditary Diseases," p. 19; Williams's "Principles of Medicine," p. 47; Paget's "Surgical Pathology," p. 514; "Cyclopædia of Anatomy and Physiology," vol. ii., p. 471; Aitken's "Science and Practice of Medicine," vol. ii., p. 35.

affections are consumption, mesenteric disease, diarrhœa, dysentery, hydrocephalus, and glandular swellings, the symptoms varying with the organs affected.[1]

Scrofulous diseases are of common occurrence in horses, cattle, sheep, and swine, either in a congenital form, or as a predisposition that may be actively developed at any period of life. In treating of the hereditary diseases of cattle, Finlay Dun remarks that "a tendency to consumption and to dysentery is often indicated by certain well-marked signs. In cattle the most obvious of these are a thin and often apparently long carcass, narrow loins and chest, flat ribs, undue length between the prominence of the ilium and the last ribs, giving a hollow appearance to the flanks, extreme thinness and fineness of the neck and withers, hollowness behind the ears, fullness under the jaws, a small and narrow muzzle, . . . hard, unyielding skin, . . . thin and dry hair, irregularity in the changing of the coat, inaptitude for fattening, prominence of the bones, especially about the haunch and tail, and want of harmony among the different parts of the body, giving the animal a coarse and ungainly look—appearances all indubitably hereditary, and indicative of a weak and vitiated constitution, and of a decided scrofulous diathesis."[2]

The peculiarities enumerated are all indications of

---

[1] For a more extended description of this class of diseases the following authorities may be consulted: *Journal of the Royal Agricultural Society*, vol. xiv., p. 124, vol. xv., pp. 79, 82, vol. xvi., p. 21, etc.; Aitken's "Science and Practice of Medicine," vol. ii., p. 215; Gross's "System of Surgery," vol. i., p. 264; and other standard medical works.

[2] *Journal of the Royal Agricultural Society*, vol. xv., p. 82.

defective nutrition, which is the essential element of the disease.

The same characters, with but little variation, are also indicative of the scrofulous habit in horses, sheep, and swine.[1]

In the heredity of scrofula, it appears that the constitutional defect is readily transmitted; but it may present itself in a different form from that observed in the generations immediately preceding. If the lungs are affected in one generation, the inherited predisposition of the next may consist in a tendency to glandular swellings, mesenteric disease, or some

[1] Scrofula in its various forms may be developed in animals that are not predisposed to it by inheritance. The most potent causes of the disease in such animals are, protracted disorder of the digestive organs, food deficient in quality and quantity, impure water, confinement in damp, filthy stables, that are not well ventilated nor lighted, exposure to cold, or any other condition that lowers the vital powers. The too common practice of crowding a large number of animals into filthy compartments, that are not well ventilated, is undoubtedly an efficient cause of the disease. According to Dr. Aitken, " the broadest fact established regarding the exciting cause of scrofula is, that the domesticated animal is more liable to scrofulous disease than the same animal in the wild state. The stabled cow, the penned sheep, the tame rabbit, the monkey, the caged lion, tiger, or elephant, are almost invariably cut off by scrofulous affections—no doubt due to deficient ventilation, and the abeyance of normal exercise of the pulmonary function " (" Practice of Medicine," vol. i., p. 234).

When the predisposition to scrofula is inherited, these conditions will be intensified in their action as exciting causes of the disease. Swine are said to be particularly liable to scrofulous affections, but this is not surprising, as the too general violation of sanitary laws that prevails in their management cannot fail to develop the disease, even in cases where the system is not predisposed to it by inheritance. Some of the most fatal epidemics among swine may be caused, in part at least, by the development of scrofulous disease.

other scrofulous affection, while the tendency to lung-disease may make its appearance in the next or some subsequent generation.

The same may be said of all so-called "constitutional diseases," the organ affected determining the character of the symptoms that indicate the presence of the general defect of the system.

When the general constitutional predisposition is inherited, the conditions to which the animal is subjected as to food, exposure, etc., may have an influence in determining the particular organ in which the disease is developed.[1]

Dr. Gross says, "The children of consumptive parents are often cut off by the same disease, or they suffer in various parts of the body, as the bones and joints, lymphatic ganglions, eye, ear, and serous membranes."[2]

In 1,000 cases of consumption tabulated by Dr. Cotton, 367 were hereditary, and of these the brothers or sisters were likewise affected in 126 cases. Of the 114 males whose parents were affected, 59 inherited the disease from the father, 40 from the mother, and 15 from both. Of the 127 females whose parents

---

[1] The injudicious use of active medicines may also be mentioned as an efficient exciting cause of the development of a disease to which the animal is predisposed, and the organs subjected to the action of such medicines will in all probability become the seat of the affection. A severe cathartic, for example, may thus develop the hereditary tendency to chronic diarrhœa or dysentery; or a profuse bloodletting may lower the general tone of the system, and thus favor the influence of other depressing agencies in developing the disease.

[2] "System of Surgery," vol. i., p. 265. *See* also Dr. Allen on "Hereditary Disease," p. 7.

were affected, 53 inherited the disease from the father, 62 from the mother, and 12 from both.[1]

A comparison of 1,031 consumptives with 1,031 non-consumptives, insured in the Mutual Life Insurance Company of New York, shows that "nearly twice as many of the former had consumptive blood-relations as of the latter, or, to speak more accurately, 18.81 per cent. of the consumptives, and only 10.89 per cent. of the non-consumptives, had near relations (parents or brothers or sisters) who died of consumption."

These "cases were all healthy lives, selected after medical examination, and one of the rules of this examination tended to exclude persons with a decided family taint; hence we should expect to find here a much smaller number of tainted families than among consumptives in general." [2]

The transmission of mental peculiarities, referred to in the preceding chapter, is not confined to those idiosyncrasies that are compatible with what may be termed a healthy condition of the nervous system, but extend also to the various forms of mental disease. Among 1,375 lunatics, Esquirol found 337 cases of hereditary transmission.[3]

In 50 cases of insanity examined by Maudsley, 16

<hr>

[1] "On the Nature, Symptoms, and Treatment of Consumption," by R. P. Cotton, M. D., London, 1852, p. 61; quoted in the *Journal of the Royal Agricultural Society*, vol. xvi., p. 35.

[2] "Mortuary Experience of the Mutual Life Insurance Company of New York," vol. ii., pp. 71–73.

[3] *Popular Science Monthly*, November, 1873, p. 58; London *Lancet*, quoted in the *Pacific Medical and Surgical Journal*, February, 1877, p. 406.

were hereditary. In 73 cases given by Trelat, 43 are represented as due to heredity. "From a report made to the French Government in 1861, it appears that, in 1,000 cases of persons of each sex admitted to asylums, 264 males and 266 females had inherited the disease. Of the 264 males, 128 inherited from the father, 110 from the mother, and 26 from both. Of the 266 females, 100 inherited from the father, 130 from the mother, and 36 from both."[1]

Dr. Hammond remarks that the hereditary tendency to insanity is shown "not only by the fact that ancestors have been insane, but that insanity in the descendants may have resulted from hysteria, epilepsy, catalepsy, or some other general nervous affection in them."[2]

Bone-spavin, curbs, ring-bone, navicular disease, and other similar affections of the bones and joints, are of frequent occurrence in the hereditary form.

Many cases are on record illustrating the heredity of this class of diseases, which we need not quote, as the fact of their transmission is familiar to every one. But a single case will be given, that came under my observation several years ago.

A mare affected with ring-bone, that unfitted her for farm-work, was kept as a breeder for several years. Her colts were quite uniform in form and color, and, as they showed no indications of the disease when two or three years old, they found ready buyers at good prices.

At the age of five or six years, however, they all

---

[1] Ribot, "Heredity," p. 131.
[2] "Diseases of the Nervous System," p. 376.

had ring-bone, to a greater or less extent, and several were entirely disabled.

In horses, strain of the back-tendons, swelled legs, grease, and roaring, are often hereditary; while a predisposition to rheumatism, malignant and non-malignant tumors, chronic cough, ophthalmia and blindness, epilepsy, and a great variety of nervous disorders, is inherited by them in common with cattle, sheep, and swine.[1]

Lucas says, "A blind beggar was the father of four sons and a daughter, all blind. Dufau, in his work on 'Blindness,' cites the cases of twenty-one persons blind from birth, or soon after, whose ancestors—father, mother, grandparents, and uncles—had some serious affection of the eyes."[2]

According to M. Trehonnais, a stallion, in France, became blind from the effects of disease, and all of his progeny had the same defect before reaching the age of three years.[3]

Dr. Dun says that "a very large number of the stock of the celebrated Irish horse Cregan have become affected by ophthalmia of the worst kind. I am told by a gentleman well acquainted with this stock that the tendency is still decidedly marked, even in the fourth and fifth generations, often appearing, and sometimes speedily causing blindness very early

---

[1] "Encyclopédie Pratique de l'Agriculteur," tome viii., p. 678. *See* also a series of articles on "Hereditary Diseases," by Finlay Dun, in the *Journal of the Royal Agricultural Society*, vols. xiv., xv., xvi.

[2] Quoted from Ribot on "Heredity," p. 40. *See* also Darwin's "Animals and Plants under Domestication," vol. ii., p. 18.

[3] "Encyclopédie Pratique de l'Agriculteur," tome viii., p. 678. *See* also "The Horse," by Youatt, p. 115.

in life, as at two or three years of age, and even before the animals have been exposed to what are considered the ordinary exciting causes of ophthalmia." [1]

M. Pauli gives the case of a family of nine children who were all born blind. Sir Henry Holland
states that four out of five children in one family became blind at the age of about twelve years, the hereditary character of the defect being confirmed by
" the existence of a family monument, long prior in
date, where a female ancestor is represented with
several children around her, the inscription recording
that all the number were blind."

" In the family of Le Compte, thirty-seven children and grandchildren became blind like himself,
and the blindness in this case occurred about the age
of seventeen or eighteen years, for three successive
generations." [2]

Dr. Dun gives the case of a stallion that, at the
age of four years, " appeared perfectly sound, and his
limbs were nearly black, well formed, and fine ; within a short time, however, they became thick and
greasy. And, although the mares to which he was
put were perfectly free from such faults, the progeny
have shown, in every case where they can be traced,
unmistakable evidence of their inheriting the greasy
diathesis of their sire. They have all been found
liable to swelled legs when they stand idle for a few

---

[1] *Journal of the Royal Agricultural Society*, vol. xiv., p. 120.

[2] The last three cases are copied from Mr. Sedgwick's paper in the
British and Foreign Medico-Chirurgical Review, April, 1861, p. 250.
The case of the Le Compte family was originally reported in the *Balti
more Medical and Physical Register*, 1809.

days; most of them have been the subjects of repeated attacks of weed; all are affected, particularly in the spring, with scurfiness of the skin of the hind extremities, and excessive itchiness, and lose at a very early age their flatness and smoothness of limb."

"The faults occur, to a greater or less degree, in all the stock of this horse by many different mares, and are distinctly traceable to the third generation."[1]

The following case of bilateral symmetry in the heredity of bony tumors, reported by Dr. Paget, is of particular interest, as it illustrates a peculiarity which is also observed in other diseases : " A boy, six years old, was in St. Bartholomew's Hospital, five years ago, who had symmetrical tumors on the lower ends of his radii, on his humeri, his scapulæ, his fifth and sixth ribs, his fibulæ, and internal malleoli. On each of these bones, on each side, he had one tumor, and the only deviations from symmetry were that he had an unmatched tumor on the ulnar side of the first phalanx of his right forefinger, and that each of the tumors on the right side was rather larger than its fellow on the left. I saw this child's father, a healthy laboring-man, forty years old, who had as many, or even more, tumors of the same kind as his son; but only a few of them were in the same positions. All these tumors had existed from his earliest childhood; they were symmetrically placed, and ceased to grow when he attained his full stature; since that time they had undergone no apparent change. None of this man's direct ancestors, nor any other of his children, had similar growths; but four cousins, one female and three male children of his mother's sisters, had as many of them as himself."[1]

The inherited predisposition to any form of disease may be derived from either or both parents, but, in the latter case, it is also likely to be intensified by being made a dominant character.[2]

The hereditary predisposition to disease may not be observed in a particular individual, but its recurrence in the offspring shows that the defect has been inherited, and likewise transmitted. In such cases the influence of favorable sanitary conditions may have been sufficient to counteract the inherited tendency in some degree, or the absence of exciting causes may have prevented its development, without interfering with the potency of its transmission to the next generation. The hereditary predisposition may thus be suspended for several generations, and then reappear with an intensity that indicates the marked persistence of the hereditary taint, even in individuals that seem to be exempt from it.

The inherited predisposition to disease, in individuals apparently free from it, may often be detected by its repeated occurrence in some collateral branches of the family. This alternation in the development of hereditary disease is observed, not in rare instances only, but so frequently that it seems to be the rule,

---

[1] "Surgical Pathology," p. 465.

[2] "Cyclopædia of Anatomy and Physiology," vol. ii., p. 471; Carpenter's "Mental Physiology," p. 369; London *Lancet*, quoted in the *Pacific Medical and Surgical Journal*, February, 1877, p. 408. (For dominant characters *see* pp. 77 and 78.)

rather than the exception, in the transmission of constitutional peculiarities.

In speaking of the heredity of cancer, Dr. Paget says : " Let it be observed, this tendency to cancerous disease is most commonly derived from a parent who is not yet manifestly cancerous ; for, most commonly, the children are born before cancer is evident in the parent ; so that, as we may say, that which is still future to the parent is transmitted potentially to the offspring. Nay, more, the tendency which exists in the parent may never become in him or her effective, although it may become effective in the offspring ; for there are cases in which a grandparent has been cancerous, and, although his or her children have not been so, the grandchildren have been. Let me repeat, the cases of hereditary cancer only illustrate the common rule of the transmission of hereditary properties, whether natural or morbid. Just as the parent, in the perfection of maturity, transmits to the offspring those conditions, in germ and rudimental substance, which shall be changed into the exact imitation of the parent's self, not only in the fullness of health, but in all the infirmities of yet future age ; so, also, even in seeming health, the same parent may communicate to the materials of the offspring the rudiments of yet future diseases ; and these rudiments must, in the case before us, be such modifications of natural compositions as, in the course of many years, shall be developed or degenerate into materials that will manifest themselves in the production of cancer." [1]

In the cases of hereditary disease already noticed,

[1] " Surgical Pathology," p. 639.

the defect in the system of the ancestors is apparently transmitted directly to the offspring, where it makes its appearance in the congenital form or as a predisposition.

Animals that are, however, free from constitutional taint, may transmit indirectly to their offspring a predisposition to certain forms of disease, through a faulty conformation or proportion of the organization, that can hardly be considered abnormal.

Animals inheriting such peculiarities of structure may remain healthy under favorable conditions; but they are liable to disease, from the effects of exposure or hard work, that would not be injurious to those with a better-proportioned organization.

According to Finlay Dun, a disproportion in the width and strength of the leg below the hock to the width and strength above the hock, predisposes to spavin; a straight hock and a short os calcis, inclining forward, gives a tendency to curbs; " round legs and small knees, to which the tendons are tightly bound, are especially subject to strains ;" while a predisposition to navicular disease is found "in horses with narrow chests, upright pasterns, and out-turned toes." [1]

" Many farm-horses, as well as others without much breeding, are remarkable for consuming large quantities of food, for soft and flabby muscular systems, and for round limbs containing an unusual proportion of cellular tissue. These characters are notoriously hereditary, of which indubitable evidence is afforded by their existence in many different indi-

[1] *Journal of the Royal Agricultural Society*, vol. xiv., p. 115.

viduals of the same stock, and their long continuance, even under the best management and most efficient systems of breeding. Such characters indicate proclivity to certain diseases, as swelled legs, weed, and grease."[1]

If the leg below the hock is disproportionately long, and the os calcis is short (giving a narrow hock), a strain of the joint, or some other form of disease, is liable to result from an amount of work that would not be severe in a limb of proper proportions.

Any marked dilatation or contraction of the blood-vessels gives a tendency to irregularities of the circulation when the work performed is severe, and a consequent predisposition to congestion or inflammation of important organs.

Like an engine with a fly-wheel that is not perfectly balanced, the animal organization of faulty proportions is enabled to perform a moderate amount of work without difficulty; but, when the machinery is taxed nearly to its full capacity, the defective adjustment becomes a source of danger, involving the integrity of other parts of the system.

This indirect transmission of a predisposition to disease, through a faulty proportion of parts, is of frequent occurrence, and it will undoubtedly explain many of the cases of disease appearing suddenly, without apparent cause, and in which an hereditary taint was not suspected, from the fact that the ancestors were not affected with the disease in any form.

This form of hereditary transmission furnishes a

[1] *Journal of the Royal Agricultural Society*, vol. xiv., p. 121.

good illustration of the importance, to the breeder, of a knowledge of all the details of structure and conformation of the animal system, and the relations of peculiarities of form to strength and constitutional vigor.

The offspring of animals that are very young, with a system immature or imperfectly developed, or of those that have had their constitution impaired by abuse or overwork, will inherit a condition of the system that predisposes to attacks of disease from slight exciting causes.

The effects may not be observed in all cases in a single generation, but, if the practice of breeding from such imperfect organizations is continued for several successive generations, the most unfavorable results may be produced.

It is stated that "*precocious* marriages are not only less fertile, but the children also which are the result of them have an increased rate of mortality."[1]

Dr. Duncan adds "the evidence of two gentlemen skilled in the breeding of lambs and calves. They say that the mortality of the young of these animals, when the mothers are immature, is much greater than when they are well grown."

"One of them says : 'Taking the first lamb from ewes at one year old has in almost every case failed to be remunerative, owing to the frequent death of the lambs.' The same may be said of young heifers, though the mortality of the offspring may not be so marked as in that of sheep."[2]

[1] Dr. Duncan, " Fecundity, Fertility, and Sterility," p. 38.
[2] Ibid., p. 390.

The same author says: "Childbearing by an immature mother is popularly held to be dangerous to the continued general health of the mother, and to prevent her complete development in size and beauty. I have no positive evidence to adduce in favor of this generally-entertained notion, which my own experience appears to me to confirm. . . . In its corroboration, however, I can adduce the ample experience of eminent breeders of the lower animals. I have had this opinion expressed to me, especially in regard to mares, cows, ewes, and bitches."[1]

Many other authorities might be cited to the same effect, were it not that the influence of early breeding in arresting the development of the mother is so often observed by intelligent breeders as to render it unnecessary.

In oviparous animals it has been observed, not only that the eggs of very young females are less in number and smaller than those produced at maturity, but that a larger proportion are not fertile, the yelk being frequently wanting or imperfect. And also, in other groups of animals, that the number of young produced at a birth is less with young mothers than with those that are fully developed.[2]

Geyelin says: "It has been ascertained that the ovarium of a fowl is composed of six hundred ovules, or eggs; therefore a hen, during the whole of her life, cannot possibly lay more eggs than six hundred, which, in a natural course, are distributed over nine years, in the following proportions:

[1] Dr. Duncan, "Fecundity, Fertility, and Sterility," p. 392.
[2] Ibid., pp. 38, 65, 70.

"First year after birth	.	.	.	.	15 to 20			
Second " " "	.	.	.	.	100 " 120			
Third " " "	.	.	.	.	120 " 135			
Fourth " " "	.	.	.	.	100 " 115			
Fifth " " "	.	.	.	.	60 " 80			
Sixth " " "	.	.	.	.	50 " 60			
Seventh " " "	.	.	.	.	35 " 40			
Eighth " " "	.	.	.	.	15 " 20			
Ninth " " "	.	.	.	.	1 " 10 " [1]			

Dr. Duncan, in summing up the results of an extended collection of statistics relating to births, shows that a similar law prevails among women. While those under twenty years of age are less fecund than those between twenty and twenty-four, a gradual increase in productiveness is made to the age of thirty years, which is the most prolific age, after which a rapid decrease in fertility takes place.[2]

The influence of diminished fecundity in young mothers upon their offspring, that necessarily inherit the same peculiarity, would tend to predispose to barrenness and sterility in the breed or family in which early breeding is frequently practised; while the defective development of the mother, arising from the same cause, would become a constitutional peculiarity in the offspring.

As the retarded development of the mother and the defective condition of the germ or egg are both the result of immaturity, and a consequent deficiency in constitutional vigor, which, as we have seen, will undoubtedly be transmitted, they must have a marked

[1] "Poultry-Breeding," p. 27.

[2] Dr. Duncan, "Fecundity, Fertility, and Sterility," as quoted in Walford's "Insurance Cyclopædia," vol. iii., p. 194.

influence in producing conditions of the system that predispose to disease.[1]

[1] In addition to authorities quoted, *see* Duckham's " Lecture on Hereford Cattle," p. 5; Youatt on "Cattle," p. 526; Youatt on "The Horse," p. 221.

CHAPTER IV.

THE habits and characteristics of animals that have been developed by the conditions in which they are placed, or the peculiar training they have received at the hands of man, appear to be transmitted from generation to generation, with nearly the same certainty and uniformity as those that characterize the original type or species from which they are descended.

Some of the most striking illustrations of this form of heredity are to be found in the transmission of the highly-artificial peculiarities that characterize the various improved breeds of animals. The tendency to lay on fat rapidly and to mature early is inherited in the best families of the Short-horns—the Devons, the Herefords, and other meat-producing breeds—while the ability to secrete an abundant supply of milk is, in like manner, perpetuated in the Ayrshires, the Jerseys, and other dairy breeds.

The certainty with which these acquired qualities are transmitted constitutes one of the most valuable peculiarities of a breed.

The American trotting-horse furnishes another illustration of the inheritance of acquired characters.

The various breeds of dogs have peculiarities that

have been developed by a long course of training, which are transmitted with a uniformity that is surprising. Young setters, pointers, and retrievers, that have never been in the field, will often "work" with as much steadiness and ability as those that have had a long experience in sporting.

In such cases, however, it will be found that the ancestors, immediate or remote, have been well trained in their special methods of hunting.

The shepherd-dog is remarkable for its sagacity and the persistence with which it carries out the wishes of its master; and it would be difficult, if not impossible, to train dogs of any other breeds to equal them in their special duties. The greyhound runs by sight, and the blood-hound by scent, and their offspring all inherit the same peculiarities.

"The curious fact was observed by Mr. Knight, that the young of a breed of springing spaniels which had been trained for several successive generations to find woodcocks seemed to know as well as the old dogs what degree of frost would drive the birds to seek their food in unfrozen springs and rills." [1]

"A new instinct has also become hereditary in a mongrel race of dogs employed by the inhabitants of the banks of the Magdalena almost exclusively in hunting the white-lipped peccary. The address of these dogs consists in restraining their ardor and attaching themselves to no individual in particular, but keeping the whole in check. Now, among these dogs some are found which, the very first time they are taken to the woods, are acquainted with this mode of

[1] Carpenter's "Mental Physiology," p. 104.

3

attack, whereas a dog of another breed starts forward at once, is surrounded by the peccaries, and, whatever may be his strength, is destroyed in a moment."[1]

"A race of dogs employed for hunting deer in the platform of Santa Fé, in Mexico, is distinguished by the peculiar mode in which they attack their game. This consists in seizing the animal by the belly and overturning it by a sudden effort, taking advantage of the moment when the body of the deer rests only upon the forelegs, the weight of the animal thus thrown being often six times that of its antagonist. Now, the dog of pure breed inherits a disposition to this kind of chase, and never attacks a deer from before while running; and even should the deer, not perceiving him, come directly upon him, the dog steps aside, and makes his assault upon the flank. On the other hand, European dogs, though of superior strength and general sagacity, are destitute of this instinct, and, for want of similar precautions, they are often killed by the deer on the spot, the cervical vertebræ being dislocated by the violence of the shock."[2]

Mr. Lewes "had a puppy taken from its mother at six weeks old, who, although never taught to 'beg' (an accomplishment his mother had been taught), spontaneously took to begging for everything he wanted, when about seven or eight months old; he would beg for food, beg to be let out of the room, and one day was found opposite a rabbit-hutch,

---

[1] "Cyclopædia of Anatomy and Physiology," vol. iv., p. 1303; Carpenter's "Comparative Physiology," p. 627.

[2] "Cyclopædia of Anatomy and Physiology," vol. iv., p. 1303.

apparently begging the rabbits to come out and play." [1]

A dog, owned by myself several years ago, inherited the same accomplishment from his mother, who had been trained to sit in an erect position and hold a stick in imitation of a soldier with a musket.

This dog was taken from his mother when but a few days old, and before it had an opportunity of learning any tricks by imitation. Without any training, when a few months old, he assumed the erect position whenever anything was wanted, and, if that did not attract attention, he would "speak," with a short bark, as his mother had been in the habit of doing.

Dr. H. B. Shank, of Lansing, informs me that a cat, owned by him, had learned to open doors that were secured with a latch, and all of her descendants inherited the same peculiarity; while another family of cats, brought up with them, did not learn the trick, although they had sufficient intelligence to ask the assistance of their more expert friends when they wanted a door opened.

Girou de Buzarringues reports the frequently-quoted case of "a man who had the habit, when in bed, of lying on his back and crossing the right leg over the left. One of his daughters had the same habit from birth, and constantly assumed that position in the cradle." [2]

Darwin reports the interesting case of a boy who "had the singular habit, when pleased, of rapidly

---

[1] Herbert Spencer, "Principles of Biology," vol. i., p. 247; Goodale, "Principles of Breeding," p. 26.

[2] Quoted from Ribot on "Heredity," p. 8.

moving his fingers parallel to each other, and, when much excited, of raising both hands, with the fingers still moving, to the sides of his face on a level with the eyes; this boy, when almost an old man, could hardly resist this trick when much pleased, but, from its absurdity, concealed it. He had eight children. Of these a girl, when pleased, at the age of four and a half years moved her fingers exactly in the same way, and, what is still odder, when much excited she raised both her hands, with her fingers still moving, to the sides of her face, in exactly the same manner as her father had done, and sometimes still continued to do when alone." [1]

The handwriting of members of the same family is said to frequently present a marked resemblance; "and it has been asserted that English boys, when taught to write in France, naturally cling to their English manner of writing." [2]

"There are families in which the special use of the left hand is hereditary. Girou mentions a family in which the father, the children, and most of the grandchildren, were left-handed. One of the latter betrayed its left-handedness from earliest infancy, nor could it be broken of the habit, though the left hand was bound and swathed." [3]

Dr. Eugene Dupuy states that "he owed to his friend Dr. Gibney the opportunity of observing a family consisting of father and mother, five children, and one grandchild.

[1] "Animals and Plants under Domestication," vol. ii., p. 15.
[2] Ibid.; Ribot on "Heredity," p. 9.
[3] Ibid., p. 38.

" The father and mother were semi-ambidextrous. All the children and the grandchild are ambidextrous to an annoying degree; all of the movements which they perform with one hand are simultaneously performed by the other hand. The girls are obliged to use only one hand when dressing themselves, or when cutting patterns, and hold the other hand down by their side, because the two hands perform the same movements at the same time, and would interfere with each other.

" Attention was called to the fact that the father of the grandchild is not semi-ambidextrous.

" Dr. Dupuy has made experiments upon these persons, and has found that, if the skin of the forearm on one side be kept well dry, and a rapidly-interrupted electrical current be used, so as only to call forth reflex actions, it is possible to induce synchronous movements in the fingers of both hands, and also muscular contraction in the lumbricales muscles of the fingers, which are too rapid to be carried on by the will."[1]

Wild animals, living on islands not often visited by man, do not fear him, but allow the closest approach without hesitation.

" When the Falkland Islands were first visited by man, the large, wolf-like dog (*Canis antarcticus*) fearlessly came to meet Byron's sailors, who, mistaking this ignorant curiosity for ferocity, ran into the water to avoid them. Even recently, a man, by holding a piece of meat in one hand and a knife in the other,

[1] " Proceedings of the American Neurological Association," in the *Virginia Medical Monthly*, August, 1877, p. 392.

could sometimes stick them at night. On an island
in the sea of Aral, when first discovered by Butakoff,
the saigak antelopes, which are 'generally very timid
and watchful, did not fly from us, but, on the contrary, looked at us with a sort of curiosity.'

"So, again, on the shores of the Mauritius, the
manatee was not, at first, in the least afraid of man,
and thus it has been in several quarters of the world
with seals and the morse. I have shown elsewhere
how slowly the native birds of several islands have
acquired and inherited a salutary dread of man; at
the Galapagos Archipelago I pushed, with the muzzle
of my gun, hawks from a branch, and held out a
pitcher of water for other birds to alight on and drink.

"Quadrupeds and birds which have seldom been
disturbed by man, dread him no more than do our
English birds, the cows, or horses, grazing in the
fields." [1]

Dr. Kidder, in his description of the "sheath-bill"
(*Chionis minor*), on Kerguelen Island, says, "When I
sat down upon a rock and kept perfectly still for a
few moments, they crowded around me like a mob of
street boys around an organ-grinder," and "all seemed
perfectly fearless and trustful." [2]

That the descendants of such animals, inheriting
the accumulated experience of their ancestors, become
wild, is shown in the instinctive dread of man exhibited by the young of the same and allied species that
are frequently brought into contact with him.   G.

---

[1] Darwin's "Animals and Plants under Domestication," vol. i., p.
33; Carpenter's "Mental Physiology," p. 90.

[2] *The Popular Science Monthly*, April, 1876, p. 661.

Leroy observes that "in districts where a sharp war is waged against the fox, the cubs, on first coming out of their earths, and before they can have acquired any experience, are more cautious, crafty, and suspicious, than are the old foxes in places where no attempt is made to trap them."

"Knight, who for sixty years devoted himself to systematic observation of this class of facts, says that during that time the habits of the English woodcock underwent great changes, and that its fear of man was considerably increased by its transmission through several generations.

"The same author discovered similar changes of habit, even in bees."[1]

The marked heredity of habits has led some modern writers to claim that the instincts of animals are but the experiences of past generations, that are accumulated and established through inheritance. Many of the most valuable characteristics of the various improved breeds of animals have been produced by the inheritance of habits of the system, arising from the conditions and treatment to which they have been subjected.

The remarkable records recently made by the American trotting-horse are the result of training and inheritance.[2]

The dairy breeds of cattle inherit a marked func-

---

[1] Ribot on "Heredity," p. 17.

[2] The first trotting-match in America was made in 1818, for a stake of $1,000, against time. It was won by a horse called Boston Blue, in the then unprecedented time of three minutes ("The Horse in America," by Herbert, vol. ii., p. 133).

tional activity of the lacteal glands, which is but a modified habit of the system.

Pritchard, in his "Natural History of Man," states that the peculiar ambling pace to which the horses bred on the table-lands of the Cordilleras are trained, has, by inheritance, resulted in a "race in which the ambling pace is natural and requires no teaching."

The Norwegian ponies, descended from animals that "have been in the habit of obeying the voice of their riders and not the bridle," are said to inherit the same peculiarity, so that it is difficult to break them to drive in the ordinary way.[1]

The habit of migration at particular seasons of the year is inherited, and I have often observed it in mallard ducks bred for several generations in a state of domestication.

It must be admitted, however, that acquired habits are not in all cases hereditary, but it would be difficult, perhaps, in the present state of our knowledge of the subject, to fix a limit to their inheritance, so far, at least, as a predisposition is concerned.

Acquired habits and the original traits of animals appear to be conflicting elements in their constitution, either one of which may, from its intensity, predominate in hereditary transmission.

Pigs have been taught to point game and to perform various tricks, but, in the hereditary transmission of their characters, "Nature" has had a stronger influence than "culture."

---

[1] The last two statements are quoted from Goodale's "Principles of Breeding," p. 25. See also "Cyclopædia of Anatomy and Physiology," vol. iv., p. 1313.

Carpenter, in discussing the heredity of acquired habits, says, " There seems to be reason to believe that such hereditary transmission is limited to acquired peculiarities which are simply *modifications* of the natural constitution of the race, and would not extend to such as may be altogether foreign to it."[1]

From a practical point of view, however, the inheritance of acquired characters, so far as they are of any value, is, fortunately, without any apparent limit.

Abnormal characters are frequently hereditary, but they are not so likely to be transmitted as acquired habits that are in harmony with the original peculiarities of the animal.

The following examples will sufficiently illustrate this form of inheritance :

Gratio Kelleia, the Maltese, " was born with six fingers upon each hand, and a like number of toes to each of his feet." He " married when he was twenty-two years of age, and, as I suppose there were no six-fingered ladies in Malta, he married an ordinary five-fingered person.

" The result of that marriage was four children : the first, Salvator, had six fingers and toes, like his father; the second was George, who had five fingers and five toes, but one of them was deformed, showing a tendency to variation ; the third was Andre—he had five fingers and five toes, quite perfect; the fourth was a girl, Marie—she had five fingers and five toes, but her thumbs were deformed, showing a tendency toward the sixth. These children grew up, and, when they came to adult years, they all married, and of course it

[1] " Mental Physiology," p. 104.

happened that they all married five-fingered and five-toed persons. Now let us see what were the results. Salvator had four children—they were two boys, a girl, and another boy—the first two boys and the girl were six-fingered and six-toed like their grandfather; the fourth boy had only five fingers and toes.

"George had only four children; there were two girls with six fingers and six toes; there was one girl with six fingers and five toes on the right side, and five fingers and five toes on the left side, so that she was half-and-half. The last, a boy, had five fingers and five toes. The third, Andre, you will recollect, was perfectly well formed, and he had many children whose hands and feet were all regularly developed.

"Marie, the last, who of course married a man who had only five fingers, had four children: the first, a boy, was born with six toes, but the other three were normal."[1]

"In a paper contributed to the *Edinburgh New Philosophical Journal*, for July, 1863, Dr. Struthers gives several cases of hereditary digital variations. Esther P——, who had six fingers on one hand, bequeathed this malformation along some lines of her descendants, for two, three, and four generations. A—— S—— inherited an extra digit on each hand and each foot, from his father; and C—— G——, who also had six fingers and six toes, had an aunt and a grandmother similarly formed."[2]

A deficiency in the number of fingers, or in the number of the phalanges or joints of the fingers and

---

[1] Huxley on "The Origin of Species," p. 92.

[2] Herbert Spencer's "Principles of Biology," vol. i., p. 243.

toes, may likewise be transmitted, as shown in the following cases from Mr. Sedgwick's paper on the " Influence of Sex in Hereditary Disease : "

A pastry-cook at Douai, named Augustin Duforet, had but two phalanges to all his fingers and toes. This defect he inherited from his grandfather, who had three children with the same malformation ; the eldest of them (a son) had three sons all with the same defect ; the second (a daughter) has had five children, two daughters with three phalanges, and three sons who have only two ; the third, who is the father of Augustin, had eleven children, five daughters normally formed, and six sons, in all of whom there is a phalanx wanting in both fingers and toes.

The mother of Augustin also had two male, still-born children, with the same deformity.

Dr. Lepine reports the case of a man who had only three fingers on each hand, and four toes on each foot ; his grandfather and son had likewise the same deformity.

Béchet records the case of a woman (Victorie Barré) " who, instead of hands, had on each arm one finger only, the other fingers and their metacarpal bones, with the exception of imperfect rudiments of two of the latter, being entirely wanting ; while on each foot there were but two toes, apparently the first and fifth, but both very defective. She was twice married : by her first marriage she had a healthy and regularly-formed male child, and by her second marriage two daughters malformed like herself ; and her sister and father were also deformed in a similar manner."

Another case is on record of the "hereditary absence of the two distal phalanges," in which "the transmission of the defect for ten generations had been effected by the females only of the family."[1]

A supernumerary organ, when inherited, may occupy a different position from that observed in the parent, as in the case of a woman with three nipples, published by Adrien de Jussieu. "The additional nipple was placed in the groin, and served ordinarily for suckling, while in the mother of this woman, who was born also with three nipples, they were all placed on the anterior region of the thorax."[2]

The fifth toe of Dorking fowls, which is one of the characteristics of the breed, has been inherited, it is claimed, from a five-toed variety introduced into Britain by the Romans. Whether this is true or not, it is now impossible to determine, but the constancy of this peculiarity, even in the produce of other breeds crossed with the Dorking, would seem to indicate that it is a character which has been fixed by long-continued inheritance.[3]

In the Houdan fowls, when first introduced into England from France, a fifth toe was rarely seen; but at the present time it is nearly as constant in this breed as in the Dorkings.[4]

Mr. Wright says : " The abnormal structure of the Dorking foot is very apt to run into still more abnor-

---

[1] *British and Foreign Medico-Chirurgical Review*, April, 1863, p. 460.

[2] Ibid., July, 1863, p. 172.

[3] Wright on "Poultry," pp. 311, 312; Darwin's "Animals and Plants under Domestication," vol. ii., p. 24.

[4] Wright on "Poultry," p. 412.

mal forms, which disqualify otherwise fine birds for the show-pen. Birds are not unfrequently produced which possess *three* back-toes, or have an extra toe high up on the leg; or, in the case of the cock, with supernumerary spurs, which have been known to grow in every possible direction."[1]

This tendency to an increase in the development of an abnormal character that has become hereditary has been observed in other cases, but we are as yet unable to present a satisfactory explanation of them. In the case of the Dorking, the practice of breeding only those birds that have the abnormal peculiarity might be expected to intensify the tendency to its production, by making it a dominant character; but, in the following case given by Dr. Struthers, it will be safe to presume that only one parent had the abnormal character, and yet we find the same tendency to its increase. " In the first generation an additional digit appeared on one hand, in the second on both hands, in the third three brothers had both hands, and one of the brothers a foot, affected; and in the fourth generation all four limbs were affected."[2]

" In a family," says Sir H. Holland, " where the father had a singular elongation of the upper eyelid, seven or eight children were born with the same deformity, two or three other children having it not."[3]

Dr. Osborne reports the case of " John Murphy,

---

[1] Wright on " Poultry," p. 331.

[2] Quoted in Darwin's " Animals and Plants under Domestication," vol. ii., p. 23.

[3] " Philosophical Transactions," 1814, p. 91; quoted in Darwin's " Animals and Plants under Domestication," vol. ii., p. 17.

aged fifty-two years, a native of County Wexford
(Ireland), who had fifteen brothers and five sisters, all
of whom possessed the family peculiarity of tortoise-
shell-colored eyes. The inheritance was derived from
the mother, whose maiden name was Murray. She
had three sisters and one brother, who were all similarly affected, and who inherited the peculiarity from
their mother, whose maiden name was F——. It is
to this latter family that the peculiarity belongs, insomuch that in the part of the country where they re-
sided they have been commonly recognized by this
distinction, and celebrated for communicating it to
their posterity." In this case, for three generations
"the transmission of the defect has been restricted
exclusively to the female sex." [1]

"In the year 1770, as we learn from D'Azara, a
hornless bull was produced in Paraguay, which has
been the progenitor of a race of hornless cattle that
has since multiplied extensively in that country." [2]

The polled breeds of Great Britain undoubtedly
had a similar origin.

According to Dr. Randall, "a ram having ears of
not more than a quarter of the usual size appeared in
a flock of Saxon sheep in Germany. He was a superior animal, and got valuable stock. These were in-
terbred, and a 'little-eared' sub-family created. Some
of these found their way into the United States, between 1824 and 1828. One of the rams came into

---

[1] *British and Foreign Medico-Chirurgical Review*, April, 1861, p.
248. The case was originally published in the *Dublin Medical Journal*
for 1835.

[2] "Cyclopædia of Anatomy and Physiology," vol. iv., p. 1311.

Onondaga County, New York. He was a choice animal, and his owner, David Ely, valued his small ears as a distinctive mark of his blood.

"He bred a flock by him, and gradually almost bred off their ears entirely.

"His flock enjoyed great celebrity and popularity in its day, but has long been broken up, and many years have doubtless elapsed since any of the surrounding sheep-owners have used a 'little-eared' ram; yet nearly every flock that retains a drop of that blood —even coarse-mutton sheep bred away from it, probably for ten or fifteen generations, insomuch that all Saxon characteristics have totally disappeared—still continues to throw out an occasional lamb as distinctly marked with the precise peculiarity under consideration as Mr. Ely's original stock."[1]

The "Ancon" or "Otter" breed of sheep, that originated in Massachusetts in 1791, were characterized by the length of their bodies and the "extreme shortness of the legs, which also turned out in such a manner as to render them rickety. They cannot run or jump, and even walk with difficulty."[2]

This deformed breed is said to be descended from

[1] "Practical Shepherd," p. 104.

[2] These sheep were described by Colonel Humphreys, in the "Philosophical Transactions," London, 1813, p. 88, according to Darwin ("Animals and Plants under Domestication," vol. i., p. 126), who states that this breed had their origin on the farm of Seth Wright, in Massachusetts.

Chancellor Livingston, in his "Essay on Sheep," 1813, p. 37, from which the description above is quoted, says, "The Otter sheep, it is said, were first discovered on some island on our Eastern coast, where, I cannot precisely say, and from thence they have spread to the adjoining States."

a ram in which the malformation was congenital. It is stated, on the authority of Colonel Humphreys, that this defect became so fixed by inheritance that it was uniformly transmitted.

The Niata cattle, on the northern bank of the Plata, described by Darwin, have a peculiar malformation of the skull, that undoubtedly has been developed by the inheritance of a deformity of some of the ancestors.

In this breed " the forehead is very short and broad, with the nasal end of the skull, together with the whole plane of the upper molar teeth, curved upward. The lower jaw projects beyond the upper, and has a corresponding upward curvature." [1]

A very singular abnormal peculiarity is hereditary in some families of pigs—the tail, which is perfectly formed at birth, having a tendency to waste away and drop off when the animals are a few weeks old.[2]

Cases are reported of families with a single lock of hair of a different color from the rest of the hair, which in one generation may be upon the right side, and in the next on the left.[3]

A family of my acquaintance have several abnormal peculiarities that are transmitted with great uniformity. The little toes lap over the adjoining toes, and the nails have a longitudinal groove that gives them a bifid termination, so that when the nail is trimmed the part cut off is in two pieces. This same character of the nail is seen also on the index-fingers.

---

[1] "Animals and Plants under Domestication," vol. i., p. 113.
[2] *Journal of the Royal Agricultural Society,* vol. xvi., p. 41.
[3] "Animals and Plants under Domestication," vol. ii., p. 14.

In addition to these peculiarities, a cartilaginous projection on the back of the ear is inherited. The paternity of an illegitimate child, in one instance, was traced to this family, from its inheritance of the peculiarities above-mentioned.

Dr. Anderson says a gentleman of his acquaintance "chanced to find a rabbit among his breed that had only one ear; he watched the progeny of that creature, and among these he found one of the opposite sex that had only one ear also; he paired these two one-eared rabbits together, and has now a breed of rabbits with one ear only, which propagate as fast, and as steadily produce their like, as the two-eared rabbits from which they originally were descended." [1]

The same author gives the case of a bitch that was born with only three legs. " She has had several litters of puppies, and among these several individuals were produced that had the same defect with herself." [2]

He also states that "a cat belonging to Dr. Coventry, of Edinburgh, which had no blemish at its birth, lost its tail by accident when it was young.

" It has had many litters of kittens, and in every one of these there was one or more of the litter that wanted the tail, either in whole or in part." [3]

" Blumenbach affirms that ' a man whose little-finger of the right hand had been nearly demolished and set awry had several sons, all of whom had the little fingers of the right hand crooked.' " [4]

[1] " Recreations in Agriculture," vol. i., p. 68.
[2] *Loc. cit.*, p. 68.   [3] Ibid., p. 69.
[4] As quoted in the *British and Foreign Medico-Chirurgical Review*, April, 1863, p. 462.

In his experiments with Guinea-pigs, Dr. Brown-Séquard observed that, in those subjected to a particular operation, involving a portion of the spinal cord or sciatic nerve, "a slight pinching of the skin of the face would throw the animals into a kind of epileptic convulsion. When these epileptic Guinea-pigs bred together, their offspring showed the same predisposition, without having been themselves subjected to any lesion whatever; while no such tendency showed itself in any of the large number of young which were bred from parents that had not been operated on." [1]

Prof. Tanner says he knew "a very striking instance of the loss of milk in a flock (previously celebrated for their supply of milk) being traced entirely to the use of a very well-formed ram, bred from a ewe singularly deficient in milk." [2]

It is stated on good authority that animals that have been "branded" in the same place for several successive generations, transmit the same mark to their offspring." [3]

From the many cases of inherited habits and abnormal peculiarities on record, we have quoted a sufficient number to show the great variety of such characters that are liable to be transmitted.

In a large proportion of cases it must be admitted that the abnormal peculiarities of parents are not observed in the offspring, and it has been claimed from

[1] Carpenter's "Mental Physiology," p. 371 ; Darwin's "Animals and Plants under Domestication," vol. ii., p. 36; Herbert Spencer's "Biology," vol. i., p. 251.

[2] *Journal of the Royal Agricultural Society*, vol. xxii., p. 5.

[3] "Encyclopédie Pratique de l'Agriculteur," tome viii., p. 678; Goodale's "Principles of Breeding," p. 25.

this fact that they have not been transmitted. From the cases presented in the following chapter, however, it will readily be seen that the non-inheritance of a character can only be determined by an exhaustive examination of the individuals in the collateral branches of the family, as well as those in the direct line of descent. If a character does not make its appearance in a particular instance, it does not necessarily follow that it has not been inherited, as it may be obscured or made latent by the presence of some other character that for the time is dominant in the organization.

The heredity of acquired habits and abnormal peculiarities should not be considered as exceptional, but rather the result of some general law of the organization that is constant in its action, and the supposed cases of non-inheritance of a character will in all probability be found to be in accordance with it.

It has been supposed that the transmission of functional peculiarities of an organ involved the transmission of some corresponding structural change of the organization, that gave rise to the abnormal modification of its function.

There are cases, however, in which a well marked functional derangement of certain organs, originally produced by an injury to the nervous system, has become hereditary, without the transmission of any apparent malformation of the nerves themselves. Dr. Eugène Dupuy has given some interesting illustrations of this singular form of heredity, some of which he observed as the assistant of Dr. Brown-Séquard, in his experiments on Guinea-pigs, already noticed, while others are the result of his own investigations.

"If in a Guinea-pig, for instance," says Dr. Dupuy, "that portion of the vaso-motor branch (of nerves) which is in connection with the carotid artery in the neck—which, therefore, regulates the blood-supply of some part of the brain, of the ear, of the face, and of the eye—be divided, or, better still, if the ganglion from which that branch springs be removed, we see that the entire half of the head of the animal, on the side on which the operation has been performed, becomes hotter, and, on examining more closely, we discover that the increase of heat is due to the fact that the blood-vessels allow more blood to pass through them, that the nutrition of the parts is increased, and therefore the heat also increases; and we see that the upper eyelid of the animal drops a little, being in a state of hyperæmia—that is, its capillaries are distended—that the secretion of tears is increased, so that the eye is wet, that the pupil of the eye is contracted, because of more blood in the ciliary system, etc.

"The ear also becomes hotter, and, if the animal is white, we can see that the ear which before was white, with some blood-vessels stretching across, is now become red, and presents a rich network of capillaries, which have become apparent, being of enlarged calibre. Now, all these phenomena may disappear after a while, except a few. The eye always remains smaller, although the blood-supply of the eyelid is more regulated; the pupil remains a little contracted and the secretion of tears continues, and also the nictitant membrane remains in a congested state. No matter how long the animal lives, that state of the eye persists, and, when the animal dies, or is sacrificed, it is seen that this eyeball is smaller than its fellow.

" If, now, such an animal were allowed to breed with another, whether operated upon in the same manner or not, it would be seen that young which are born apparently perfectly healthy present, a few days after birth, all the phenomena observed in their changed parent or parents. They have the same smaller eyes, but on both sides, the same ear thickened and enlarged, etc.

" The only phenomena which they do not show are those which have been transient—the increased heat and the increased sensation which depended upon the increased amount of blood present, etc. Those young can be made to breed in-and-in for several generations. I have watched them for five generations, and always the same characteristics will be discovered in the young."

" If, now, an examination is made of the parent, the first one, it will be seen that the nerve that had been sectioned, or its ganglion which had been extirpated, is not regenerated; while, if an autopsy is made of one of the offspring of any of the subsequent generations, it is seen that they all possess the nerve and the ganglion intact. The acutest or most minute microscopic examinations do not discover any difference between their structure and those of other animals of the same family and species." [1]

In these cases the permanent modifications of the eye and face, resulting from the injury to the nervous system, are entailed upon the offspring, while the

[1] *Popular Science Monthly,* July, 1877, pp. 333, 334.

nerves that have been mutilated are transmitted in their original integrity.

The following cases, given by Dr. Dupuy, are of particular interest from the series of changes repeated in the offspring that have not apparently inherited the original lesion of the nervous system that produced them :

" If a puncture be made into that portion of the upper part of the spinal cord which anatomists call the restiform body, in Guinea-pigs, it will be seen that the animal presents at once an increased vascularity of the ear on the corresponding side ; the ear becomes gorged with blood, chiefly toward the periphery ; sometimes, in a very short time, indeed, that portion of the ear falls off, destroyed by dry gangrene.

" I have the record of a case in which the ear was thus partially destroyed in less than nine hours. The eye on the same side becomes larger and protrudes ; it protrudes first, and becomes larger in the course of time. If a pair of Guinea-pigs thus operated upon be allowed to breed, and even if only one parent is thus diseased, the other being healthy, when young are born these young always present the phenomena observed in the parents ; but the phenomena just described only come shortly after their birth.

" It is seen that their eyeballs increase in size and protrude from their sockets ; their ears after a few days become diseased, just like those of the parents, the subjects of experimentation, and drop off, eaten by dry gangrene.

" When the parent or parents are sacrificed, and their restiform bodies are examined microscopically, nothing is detected but a cicatrix in the envelopes of the spinal cord, which appears a little thickened at that point; but the nervous tissue itself does not differ apparently from surrounding elements of the same nature and structure.

"If an examination is also made of one of the young, nothing at all is discovered.

"These young can be allowed to breed in-and-in, and always the same phenomena will be observed in each subsequent generation.

"I have sometimes noticed that if a male or a female belonging to any one of the successive generations is allowed to breed with another healthy animal, very generally some of the young present the same hereditary peculiarities. I have followed animals thus operated upon through seven generations." [1]

In the experiments of Dr. Brown-Séquard with Guinea-pigs, it was found that an injury of the spinal cord, or of the sciatic nerve, produced a change in sensation over a certain well-defined area of the face, in addition to the epileptic affection already referred to. When the sciatic nerve was the seat of the injury the outer part of the foot was likewise destroyed, leaving but one toe, the inner, on the foot of the injured side, and this deformity is a permanent one.

When the animals recover from the epileptic affection, as they do after several months, "all the phenomena observed about the zone of skin in the neck and face recur in the reverse order; that is to say, all the different sensations return by degrees, at the same time that the hair of the region falls, and new hair

[1] *Popular Science Monthly*, July, 1877, pp. 334, 335.

grows gradually. The fits become simple convulsions,
then mere twitchings, and lastly the animal can no
longer be distinguished from another healthy one, but
by the fact that it has only one toe at one of its hind-
legs, when the operation has been performed on the
sciatic nerve; and nothing whatever remains when
the origin of the disease was a prick in the spinal
cord."[1]

The young of these epileptic Guinea-pigs are born
apparently healthy, with the exception of those from
parents that had been subjected to the injury of the
sciatic nerve, and they have but one toe on one of the
hind-feet. When these apparently healthy animals
are two or more months old they gradually become
affected with epilepsy, and the same area on the face
and neck passes through the same series of changes in
the development and cure of the affection that had
been observed in their parents. " We see the gradual
increase of the affection, the diminution of the sensi-
bility in the zone, just as with the parents, the coming
of a period of complete attacks of epilepsy, and then
the loss of hair and the gradual diminution of the
nervous complaint."[2]

In the original parents, it will be observed that the
derangement of the nervous system, resulting in convulsions, was produced by an injury to the spinal cord
or the sciatic nerve, and, when these injuries had
healed, the nervous symptoms gradually disappeared,
the hair is shed from that part of the face affected, and
gradually replaced, and the cure is complete.

[1] *Popular Science Monthly*, July, 1877, p. 337.
[2] *Loc. cit.*, p. 337.

Now, the young of these animals that had recovered from their injuries are born with a nervous system that is apparently perfect; and yet, after a time, the disease is developed, passes through its peculiar stages without apparent cause, and finally disappears.

The functional derangement of the organization is apparently transmitted without being accompanied by any anatomical lesions that can be assigned as an exciting cause.

4

CHAPTER V.

ANY peculiarity of an ancestor, more or less remote, whether of form, color, habits, mental traits, or predisposition to disease, may make its appearance in the offspring without having been observed in the parents.

This form of heredity, technically termed *atavism* (from *atavus*, an ancestor), is called reversion by Mr. Darwin, and it has for a long time been recognized by breeders, under a variety of names, as " throwing back," " crying back," " breeding back," etc.

It will, perhaps, be better to retain the term atavism, which has been so generally in use to indicate this class of cases, as it does not involve in its signification any theoretical explanation of the phenomena.

Some of the cases cited in the preceding pages, to illustrate other phases of the great law of heredity, are likewise examples of atavism, and we shall find also in the cases quoted in this connection many illustrations of topics discussed in other chapters. Of the multitude of cases on record of this form of heredity, the following will serve to illustrate its leading features. Mr. Darwin states that the following case was communicated to him on good authority : " A pointer-bitch produced seven puppies. Four were marked with blue and white, which is so unusual a color with pointers that she was thought to have played false with one of the greyhounds, and the whole litter was condemned ; but the game-keeper was permitted to save one as a curiosity.

" Two years afterward a friend of the owner saw the young dog, and declared that he was the image of his old pointer-bitch Sappho, the only blue-and-white pointer of pure descent which he had ever seen. This led to close inquiry, and it was proved that he was the great-great-grandson of Sappho ; so that, according to the common expression, he had only one-sixteenth of her blood in his veins." [1]

Mr. Tollett, of Betley Hall, crossed his fowls with Malays, and, though he attempted to get rid of this strain, he gave it up in despair, the Malay characters reappearing forty years after the cross was made.[2]

Mr. Hewett states that the Rumpless fowls in some instances produce young with tail-feathers, but that, when three such birds were selected to breed from, there was but one chick with a tail out of over twenty bred from the trio.[3]

Goodale relates an interesting case that occurred in the Kennebec Valley. Many years ago there were a few polled cattle in that locality, but they finally became extinct. For thirty-five years after the last of these polled cattle was killed the cattle on the farm of Mr. Wingate all had horns, but, at the end of that

---

[1] " Animals and Plants under Domestication," vol. ii., p. 46.
[2] On the authority of Mr. Darwin, *loc. cit.*, p. 49.
[3] Tegetmeier's " Poultry-Book," p. 231.

time, a polled animal made its appearance in his herd, with all the characteristics of the original breed.[1]

It is stated, on the authority of Mr. Sidney, that in a litter of Essex pigs two young ones appeared with marks of the Berkshire that had been used as a cross twenty-eight years before.[2]

The occasional appearance of horns in the Galloway, Suffolk, and other polled breeds that have been bred pure for many years, furnishes an illustration of the transmission of an original character by atavic descent.

Mr. Sedgwick says, " In the well-known case of George III., the insanity was transmitted in the male line, by atavic descent from a male ancestor, eight generations back, in whom not only the insanity, but many other of the well-known characteristics of the unfortunate monarch, were *exactly* repeated."[3]

In the case of a woman with a sixth finger on one hand, related by Dr. Struthers, only one out of eighteen children had an extra finger, and, in this case, both hands were affected. One of the sons, James, had two sons and seven daughters, all, like himself, with the normal number of fingers. One of his daughters, however, had a son with six fingers on each hand.

Two generations were thus free from the defect, but, when it made its appearance in the next genera-

---

[1] "Principles of Breeding," p. 65.

[2] "Animals and Plants under Domestication," vol. ii., p. 49; from Youatt on "The Hog," 1860, p. 27.

[3] *British and Foreign Medico-Chirurgical Review*, April, 1863, p. 467. *See* also " The Four Georges," by Thackeray, pp. 5, 6, 1861.

tion, the intensity of transmission was increased rather than diminished, as both hands were affected instead of one, as in the case of the great-great-grandmother.[1]

Dr. Chadbourne reports a case that came under his own observation, of two young men who were cousins, " each of whom had six toes upon his feet."

Neither of the parents had the defect, but it was a characteristic of the grandparents, and appeared in the family a long time before.[2]

Mr. Sedgwick, in his article on the " Influence of Sex in Hereditary Disease," says, " Siebold records the case of a married couple whose fathers were both red-headed, but not having red hair themselves, who had four sons red-headed, and three daughters whose hair was of another color." [3]

In the Short-horn herd-books may be found numerous instances of the atavic inheritance of color, and almost every breeder can furnish from his own experience many cases of a similar character. The following is cited as an example of this class of cases : " Mr. Wadsworth owns the twin Princess cows, Lady Mary seventh and eighth ; they are both good roans, got by fourth Lord of Oxford (5903 " American Herd-Book "), a roan bull ; their dam, Lady Mary, a red, got by Hotspur (31393), a roan ; their granddam, Baroness, a red roan, got by Barrington (30501), a white ; their great-granddam, the imported red Princess cow Red Rose second, got by Napier (6238), red roan. These twin heifers, Lady Mary seventh and eighth, were both

---

[1] Spencer's " Principles of Biology," vol. i., p. 258.

[2] " Agricultural Report of Massachusetts," 1866–'67, p. 88.

[3] *British and Foreign Medico-Chirurgical Review*, April, 1863, p. 451.

served by the Princess bull, Earl of Seaham (8077 "American Herd-Book"), a good roan, and each dropped a bull-calf; but the one from Lady Mary seventh was a *red*, while the other, from Lady Mary eighth, was *white*."[1]

The late Hon. Charles Rich, of Lapeer, who, when a young man, had charge of the merino sheep that formed the foundation of what is known as the "Rich family of Merinoes," informed me that tan-colored ears was a common characteristic of the Spanish merino sheep at that time, and that it was highly prized as an indication of the "blood." Dr. Randall says: "These spots were highly characteristic of several of the families of merinoes originally imported from Spain, and the lambs of some of them were occasionally covered over the carcass at birth with larger spots of the same color, or of a deeper tawny-red. Sometimes the whole body was thus colored. But all of these tints disappeared on the body when the wool grew out, and were seen no more."[2]

These tan-colored spots on the ears and face, and also on the body, are now frequently seen in flocks in which white ears have been the prevailing characteristic for many generations, the original peculiarities of the breed being transmitted by atavism.

The "dark noses," so frequently seen in shorthorns, are but a repetition of ancestral characteristics by atavic descent.

The following case of atavic transmission of an abnormal peculiarity is reported by Mr. Sedgwick, on

[1] "The Country Gentleman," 1876, p. 105.
[2] "The Practical Shepherd," p. 72, note.

the authority of Dr. Cotton : " A gentleman had, with both dentitions, a double-tooth in place of the left-second incisor in the upper jaw ; he was the only one in the family of nine children who presented this peculiarity, which he inherited from his paternal grandfather, whom he so exactly resembled, even in the form of the hands also, as often to have arrested the attention of their acquaintance."[1]

The same authority says : " Borelli, quoted by Rougemont, records the case of a well-made man who was three times married, and whose father had been lame ; the children of this man by his three wives were all lame."[2]

The following case of skin-disease (*ichthyosis*), reported by Mr. Sedgwick, illustrates a singular feature in the atavic transmission of disease, from the limit of the defect to the male sex, while its transmission appears to be exclusively limited to females : " It first occurred in the grandfather, who is still living, and who has the disease in a very severe form ; it did not appear in him, or it was not, at least, noticed, till he was about seven or eight years old.

" This man has had three sons and three daughters. One son died at the age of five years, and one at the age of seven years, both of whom were free from the disease. The other son is living and past middle age, but has shown no tendency to the disease. The three daughters have all lived to grow up and marry, and in them likewise the skin is unaffected. Two only of the three daughters have had children. The eldest daughter has had four, of whom the first-born, a girl,

[1] *Loc. cit.*, April, 1863, p. 454.   [2] Ibid., p. 464.

has had no appearance of the disease; the three other children are boys, of whom the eldest, aged fourteen years, and the youngest, aged nine years, suffer from the disease, while the other son, aged eleven years, is free from it.

"The family of the other daughter consists of three children, the eldest of whom, aged six years, is, as in the former case, a girl, and free from the disease, while the two other children, who are boys, aged respectively three years and one year, have the skin very decidedly affected. It is to be noted that the disease, in these grandchildren, has in each case appeared within a few months after birth."[1]

A tendency to excessive hæmorrhage, from even slight injuries, is well known to be hereditary, and this to such an extent that "in some families scarcely a single male arrives at maturity." In his remarks on the heredity of this diseased condition of the system, Mr. Sedgwick says: "In some of these cases it is recorded that, while the males alone have suffered from the disease, the females alone have been able to transmit it, as in the case of Mr. Appleton, whose daughters conveyed the complaint to his grandsons, and who, in their turn, transmitted it through their daughters to their grandsons; the males in this family, as in many others similarly affected, never inheriting the disease direct from their fathers, but always through females from their grandfathers, as occurred in my case of ichthyosis."[2]

---

[1] *British and Foreign Medico-Chirurgical Review*, 1861, p. 246.

[2] *Loc. cit.*, July, 1861, p. 146. In the case of Mr. Appleton, above referred to, references are made to the *New England Journal of Medi-*

The tendency to an alternation of generations in the inheritance of disease, which has already been noticed, appears to be analogous in character to the alternations determined by the limitation of defects to one sex, while the other sex alone seems capable of transmitting them.

In many of the lower animals the alternation of generations is the fixed law of generation.

In the aphides (plant-lice), for example, nine or ten generations of individuals are produced in succession before those having sexual organs and capable of producing eggs make their appearance; and this succession of non-sexual generations is uniformly repeated.

The phenomena of atavism has been claimed to be but a reversion of the organization to characters belonging to an original ancestor or type.

This, in many instances, appears to be the case; but, in the alternations that have been observed in the hereditary transmission of disease, and even of normal peculiarities, the theory of reversion is far from satisfactory.

In the case of Rumpless fowls, as stated by Mr. Hewett, individuals with tail-feathers are of frequent occurrence, and these, as a rule, produce tailless progeny.[1]

If, in the case of individuals with tails, there is reversion to the original type, in those without tails,

cine and Surgery, vol. ii., pp. 221–225, 1813; Edinburgh Medical and Surgical Journal, vol. xxxvi., pp. 317–320, 1831; and vol. lxxvii., pp. 1–10, 1852.

[1] Tegetmeier's "Poultry-Book," p. 231.

bred from parents with tails, there must be reversion
again to the tailless form.   From the facts, as now
understood, it appears that two antagonistic characters
are alike inherited, either one of which may become
dominant in the offspring.

The alternation of the character, in different generations, may thus be produced by the development
of the one or the other of two characters belonging
as strictly to the organization, through inheritance, as
any other part of the system.

Although we may not be able, in the following
cases, to trace the principle of alternation in atavic
descent, they are, nevertheless, of interest in this connection, from their close resemblance, in some respects,
to the cases under discussion.

" A physician at Marseilles relates a case in which
deafness from birth occurred in three children alternately in a family of six.   The parents were not affected. . . . M. Saissy refers to a family living at Aix,
in Savoy, composed of seven children : the eldest is
deaf and dumb, the second hears perfectly, the third
is deaf and dumb, and the fourth enjoys the same advantage as the second; the fifth, sixth, and last, are
completely deaf—the last but one (the sixth) in this
case being an idiot.   There was no defect in either
parent. . . . A similar case occurs in the commune
of Bessenay, department of the Rhône; in a family
composed of eight children four are deaf and dumb,
and alternate with four who enjoy the sense of hearing."   .

Claude relates the case " of a woman who gave
birth to eight children of one and the other sex, the

first, third, fifth, and seventh, of whom attained the ordinary size, while the other four were dwarfs."

In a family of eight children, four sons alternating with four daughters, the sons were all healthy, while the daughters were all affected with brain-disease (hydrocephalus), the only one living being an infant under treatment.[1]

From the persistent appearance of the defects in these cases in regular alternate succession, we must admit the probability, at least, of the existence of some hereditary taint of the system, derived from ancestors whose history we are unable to trace.

In the chapter on "Sex" may be found cases in which the defect is limited to one sex; and this, in families of both sexes, would result in an alternation more or less regular in its inheritance.

In a large family we seldom find all of the children resembling either the father or the mother, and, in many instances, the resemblance to a grandparent or some more remote ancestor prevails to so great an extent that the obvious peculiarities of the immediate parents are obscured. Prof. Agassiz[2] has remarked that " the offspring is not the offspring of father and mother, but of grandparents as well," and he might also have included all of the ancestors in the parental enumeration.

The alternations observed in the transmission of ancestral characters, and the resemblance of offspring

---

[1] The last five cases are quoted from Mr. Sedgwick's paper on the "Sexual Limitation of Hereditary Disease," in the *British and Foreign Medico-Chirurgical Review*, July, 1861, pp. 141, 142, 146.

[2] "Agricultural Report of Massachusetts," 1866-'67, p. 82.

to a remote ancestor, that differs in many respects from the parents, cannot be referred to a "spontaneous variation" in the law of inheritance, for we cannot conceive of an effect without an efficient cause.

The repetition of some preëxisting character is so uniformly observed in all cases of apparent variation in the transmission of qualities, in which the history of the ancestors can be traced, that we cannot avoid the conclusion that these peculiarities in the heredity of the organization are the result of some constant and definite physiological law.

If the form in which the physiological units or elements of the organization were transmitted could be determined, the obscurity involved in this class of cases would in great measure disappear.

In discussing the subject of inherited resemblance, Dr. Carpenter remarks that "the question seems to have been entirely ignored, whether the union of two different natures may not produce—as in the combination of an acid and a base—a resultant essentially dissimilar to either of them."[1]

If two characters may thus blend to form a new character essentially different, there could be no constancy in the transmission of ancestral forms from generation to generation, and a wide variation from the family type would necessarily result. There could be no uniformity in the leading characteristics of our improved breeds, and, with our present knowledge of physiological science, the breeding of animals would be attended with the greatest uncertainty, from our

[1] "Mental Physiology," p. 369.

inability to predict what a given combination would produce.

Moreover, the phenomena of atavism cannot be reconciled with this hypothesis, without the further supposition that the elements of the organization, combining to form a new compound, may be again resolved into their original constituents.

When characters that have remained latent for several generations make their appearance again, with all the peculiarities that formerly distinguished them, it does not seem probable that they have passed through a series of transformations in the formation of new characters, and, at the same time, retained their original constitution.

From the facts of heredity already presented in the cases cited, it must be evident that the sum of the characters or physiological units that enter into the organization of the animal cannot be represented in the external peculiarities that alone are obvious to the senses. It is well known to breeders that many of the most important characteristics of the organization, in a given case, may not appear upon the surface, or in the functional activities of the system, and that they can only be traced in the ancestral history, and in the inherited peculiarities of offspring.

In the further discussion of these peculiar forms of heredity, it will be necessary to distinguish between the more obvious and prominent characters of the animal and those obscure characters that can only be shown to exist by their hereditary transmission to offspring. The former may be termed *dominant* characters, and the latter obscured or *latent* characters.

For many years I have been inclined to the belief
that all characters are directly transmitted as physiological units or elements of the organization, some of
which may be dominant, and thus determine the obvious characteristics of the animal, while others remain latent until they are transmitted to offspring in
which favorable conditions lead to their development,
when they, in their turn, may become dominant, and
thus obscure other characters.

That characters are transmitted in their integrity,
without transformation into other characters, is clearly
asserted by Herbert Spencer, who says, " There must
arise not an homogeneous mean between two parents,
but a mixture of organs, some of which mainly follow
the one parent, and some the other." [1]

The last clause of this statement cannot, however,
be literally accepted as a law of inheritance, as we
have already seen that the dominant characters, in a
given case, may be inherited from some remote ancestor, while the dominant characters of the parents may
become latent.

Mr. Sedgwick, in his paper on " Hereditary Disease," says : " It may be observed that in the offspring
of two dissimilar parents there is never, as a rule,
complete fusion of the two parents, but a distribution
of the characters peculiar to each ; and although this
is less strongly remarked in the offspring of the human
race than it is in that of the lower animals—as, for example, in the case of some hermaphrodite insects, in
which the family quarterings may result from specific
distinctions of sex being associated without fusion in

[1] " Principles of Biology," vol. i., p. 267.

the same specimen, yet, as regards the inheritance of disease, it will be found that the morbid characteristics of one or the other parent are either completely repeated or completely absent, but not fused together in the offspring. This is what is meant in inheritance by the doctrine of 'election,' which is based on the observation that certain attributes of organization peculiar to one parent are repeated in the offspring; and it offers a reasonable explanation of the fact that children often inherit the defects of one parent, while in many other respects they resemble the other; and the inheritance in these cases, both natural and morbid, may sometimes be conveyed to them by atavic descent." [1]

If it is admitted that the animal inherits an assemblage of peculiarities representing the aggregate of parental characters, it must follow that all of the characters of all ancestors are in like manner inherited, as each generation would inherit and transmit the peculiarities of the preceding generation, and this, in turn, would inherit and transmit the peculiarities of the next preceding, and so on indefinitely. The phenomena of atavism seem to show that we cannot set a limit to the inheritance of characters. Theoretically, a defect or peculiarity may be "bred out," as it is

<hr>

[1] *British and Foreign Medico-Chirurgical Review*, July, 1863, pp. 190, 191. As an illustration of the distinct inheritance of qualities, the case is given of " the scarce egger-moth, observed by Mr. Westwood" ("Entomologist's Text-Book," p. 397, 1838), "at Berlin, in which the front-part of the body and front-half of the wings were half male and half female, and the hind-part and hind-wings half female and half male, the characters of the male and female insect being exhibited on opposite quarters of this specimen."

termed, until it is represented mathematically by a
fraction so small as to scarcely merit attention, and
yet, as frequently observed, it may again appear in a
manner indicating that it has been constantly transmitted, without change, through a long series of generations.

Mr. Sedgwick remarks, in regard to atavism in
disease, that "no fixed boundaries, recognizable by
us, can be expected to limit its operation, for, like
other general laws in Nature, unity in principle coex-
ists with variety in results; and it is chiefly because
we are less familiar with the results of atavism in disease than we are with many other reproductive phenomena, as, for the sake of illustration, with memory,
that we hesitate to accept them, although they are
not, in themselves, more exceptional or peculiar than
some of those are which we not only never hesitate to
accept, but with which this phenomenon in morbid
development seems to be closely allied. For atavism
in disease appears to be but an instance of memory in
reproduction, as imitation is expressed in direct descent; and in the same way that memory never, as it
were, dies out, but in some state always exists, so the
previous existence of some peculiarity in organization
may likewise be regarded as never absolutely lost in
succeeding generations, except by extinction of race."[1]

It has been remarked that no two animals are precisely alike, in all details of the organization, no matter how close the relationship or how striking the resemblance; and, in connection with this, it has been
observed that instances occur in which individuals

[1] *British and Foreign Medico-Chirurgical Review*, July, 1863, p. 197.

present an assemblage of characters quite different from those that characterize the parents. These have been explained on the supposition that there must be a law of "spontaneity" which is antagonistic to that of heredity, or that the law of heredity is not constant in its action, but limited by numerous exceptions.[1]

The view we have presented of the law of inheritance would seem to preclude the necessity of any such hypothesis to account for the individual variations referred to. Many of the cases of supposed variation are fully explained on the principle of atavic descent, which is, as we have seen, but a phase of the great law of heredity.

If characters are transmitted as physiological units, it will be readily seen that, although an animal may be composed of precisely the same elements as its ancestors, the dominance of some of these, or the arrangement of the elements themselves, must give rise to individual peculiarities, or even to forms not precisely identical with those exhibited in the dominant characters of any ancestor. Any observed variations in the inheritance of form, color, or general characteristics, may thus be readily accounted for, within the limits of the characters belonging to the ancestors.

In these cases of apparent variation, the similarity of the offspring to its ancestors consists in the possession of the same assemblage of characters which is often shown in a general rather than a special resemblance. From the complexity of the elements transmitted from generation to generation, we cannot expect the offspring, in a particular case, to be the exact

[1] Ribot on "Heredity," p. 191, etc.

counterpart, in dominant characters, of either parent or of any ancestor; but, on the other hand, we have no reason to believe that any characters will appear that have not been derived by direct or interrupted descent from some ancestor.

When speaking of the resemblance of offspring to ancestors, in a popular sense, the dominant characters are alone referred to; but, as these, as has been shown, may constitute but a small proportion of the elements of the organization, a strict comparison of resemblances must include a wider range of characteristics.

In this connection, the importance of a full record of the pedigrees of breeding animals will be readily suggested, as a means of tracing the history of ancestors, for the purpose of determining the characters that are liable to be transmitted by atavic descent. As the subject of pedigree, however, involves a number of questions that have not as yet been examined, a full consideration of its practical bearings must be, for the present, omitted.

CHAPTER VI.

THE external form and general characteristics of an individual, as determined by heredity, are the result, as we have seen, of the prominence of those characters that are made dominant, and the suppression of others which, for the time being, are said to be latent. In the arrangement of these dominant characters in the organization, a principle of development and suppression appears to prevail, which is recognized by naturalists as the law of correlation. This law may be defined in general terms as follows: Any peculiarity in the development of one organ, or set of organs, is usually accompanied by a corresponding modification or suppression of organs belonging to some other part of the system. In this place we shall only notice the relations of this law to heredity, reserving for another chapter its applications in determining internal qualities from peculiarities of external conformation.

The correlated structure of animals enables the comparative anatomist, from the examination of a single tooth, or fragment of bone, to determine not only the class and order to which an animal belongs, but its habits and mode of life, and the character of the food required for its support. The celebrated naturalist, Milne-Edwards, in his article on crustacea, says, " It has long been admitted as an axiom in animal physics that, when any particular part of the body acquires a very high degree of development, certain other parts stop short of their ordinary state of evolution, as if the former had obtained their unusual increment at the cost of the latter." [1]

Cuvier, the great comparative anatomist, claimed that "all organized beings, in their structure, form a complete system, of which the parts mutually correspond and conduce to the same definite action by a reciprocal reaction. Each of these parts cannot be changed without the others changing also; and, by consequence, each of these taken separately indicates and gives all the rest." [2]

Prof. Owen, in his valuable work on the " Comparative Anatomy of the Vertebrates," gives the following illustrations of this law of development : " As vertebrates rise in the scale, and the adaptive principle predominates, the law of correlation, as enunciated by Cuvier, becomes more operative. In the jaws of the lion, e. g., there are large laniaries, or canines, formed to pierce, lacerate, and retain its prey. . . . There are also compressed, trenchant, flesh-cutting teeth, which play upon each other like scissor-blades in the movement of the lower upon the upper jaw. The lower jaw is short and strong ; it articulates to the skull by a transversely-extended convexity, or condyle, received

---

[1] "Cyclopædia of Anatomy and Physiology," vol. i., p. 757.
[2] As quoted by Prof. Owen, " Comparative Anatomy of the Vertebrates," vol. i., p. 27.

into a corresponding concavity, forming a closely-fitting joint, which gives a firm attachment to the jaw, but almost restricts it to the movements of opening and closing the mouth. The jaw of the carnivora develops a plate of bone, of breadth and height adequate for the implantation of muscles, with power to inflict a deadly bite.

"These muscles require a large extent of surface for their origin from the cranium, with concomitant strength and curvature of the zygomatic arch, and are associated with a strong occipital crest and lofty dorsal spines, for vigorous uplifting and retraction of the head when the prey has been griped.

"The limbs are armed with short claws, and endued with the requisite power, extent, and freedom of motion, for the wielding of these weapons. These and other structures of the highly-organized carnivora are so coördinated as to justify Cuvier in asserting that 'the form of the tooth gives that of the condyle, of the blade-bone, and of the claws, just as the equation of a curve evolves all its properties, and exactly as, in taking each property by itself as the base of a particular equation, one discovers both the ordinary equation and all its properties, so the claw, the blade-bone, the condyle, the femur, and all the other bones individually, give the teeth, or are given thereby reciprocally, and, in commencing by any of these, whoever possesses rationally the laws of the organic economy will be able to reconstruct the entire animal.'"

"The law of correlation receives as striking illustrations from the structure of the herbivorous mammals." A limb terminating in a hoof serves for locomotion only; it cannot be used as an organ of prehension, to grasp, seize, or tear. The ruminant hoofed animals all have a cloven hoof, and they are the only ones with horns on the frontal bone. When the hoofs are in one or two pairs, the horns are also in one or two pairs. The horned ungulates, with three hoofs, have either one horn, or two horns placed one before the other, in the middle of the skull.[1]

In the ruminants there is, moreover, a marked correlation in the form of the teeth, the articulation of the jaw, which provides for a free lateral motion in grinding their food, and the complex structure of the digestive organs.

Dr. Carpenter says : " It is perfectly true that, in a great majority of cases, the *extraordinary development* of one organ is *accompanied* by a corresponding deficiency of development in another. Thus, in the human cranium, the elements which form the covering or protection of the brain are very largely developed, while those which constitute the face are comparatively small. In the long-snouted herbivorous mammals, and in reptiles and fishes, on the other hand, the great development of the bones of the face is coincident with a very small capacity of the cerebral cavity.

" In the bat, while the anterior extremity is widely extended, so as to afford the animal the means of rising in the air, the posterior is very much lightened, so as not to impede its flight. In the kangaroo, on the other hand, the posterior members are very large and

---

[1] Owen's "Comparative Anatomy of the Vertebrates," vol. i., pp. xxvii., xxviii.

powerful, enabling the animal to take long leaps, while the fore-paws are proportionally small."[1]

In blind persons the sense of touch attains a delicacy that is surprising.

"It is well known that Dr. Saunderson, the celebrated blind Professor of Mathematics at Cambridge, not only acquired a very accurate knowledge of medals, but could even distinguish genuine medals from imitations, more certainly than most connoisseurs in full possession of their senses."[2]

Cases are on record of blind persons who could not only distinguish colors, but shades of the same color. The muscular sense which is employed by the blind, in connection with touch, in discriminating the form, peculiarities of surface, and size of objects, becomes in these cases remarkably developed.[3]

It is stated that persons affected with color-blindness frequently have a defective musical ear.[4]

The sense of smell, in some blind persons, is so exceedingly acute that they are enabled, by it alone, to recognize persons not in immediate contact with them.

"In the well-known case of James Mitchell, who was deaf, blind, and dumb, from his birth, it was the principal means by which he distinguished persons, and enabled him at once to perceive the entrance of a stranger."[5]

---

[1] "Comparative Physiology," p. 130.

[2] "Cyclopædia of Anatomy and Physiology," vol. iv., p. 1178.

[3] Ibid., *loc. cit.*

[4] *See* Dr. Earle's article in the *American Journal of the Medical Sciences*, vol. xxxv., p. 347; and article "Vision," "Cyclopædia of Anatomy and Physiology," vol. iv., p. 1453.

[5] "Cyclopædia of Anatomy and Physiology," vol. iv., p. 702.

Mr. Darwin states that "black dogs, with tan-colored feet, whatever breed they may belong to, almost invariably have a tan-colored spot on the upper and inner corners of each eye, and their lips are generally thus colored."[1]

According to the same author, "white cats, if they have blue eyes, are almost always deaf."

In the cases cited it is shown that, if there is the "least speck of color on their fur," or if even but one eye is not blue, the sense of hearing is not lost; and, in one instance, in which the iris at the end of four months began to grow "dark-colored," the cat then began to hear.[2]

It has been remarked that a white spot or blaze on the face of a horse is usually accompanied by white feet.

In the deer tribe, Prof. Baird notices a singular correlation of the horns and organs of reproduction. He says : " In all deer, except, perhaps, the reindeer, if the male be castrated when the horns are in a state of perfection, these will never be shed ; if the operation be performed when the head is bare, they will never be reproduced ; and, if done when the secretion is going on, a stunted, ill-formed, permanent horn is the result."[3]

Mr. Youatt remarks that a " multiplicity of horns is not found in any breed (of sheep) intrinsically of much value. It is generally accompanied by great length and coarseness of fleece, and which, in the ma-

---

[1] "Animals and Plants under Domestication," vol. i., p. 42.
[2] *Loc. cit.*, vol. ii., p. 396.
[3] "Patent-Office Report," Part II., "Agriculture," 1851, p. 111.

jority of these cases, assumes more the form of hair than of wool."[1]

The tusks, which attain a great size in the boar, are not fully developed in swine that are castrated.[2]

What are called the secondary sexual characters of the male are not developed in animals that are castrated; and, among birds, it has been observed that females incapable of breeding, from age or the effects of disease, sometimes assume the plumage and voice of the male.[3]

The cock of the Sebright bantams should be hen-tailed and without sickle-feathers, thus presenting a close resemblance to the female. This character, so highly prized by exhibitors, has, however, its disadvantages.

Mr. Hewitt remarks, in regard to these breeds: "The combined experience of many other admirers of the Sebright bantams is concurrent with my own, viz., that even a very trifling disposition to sickle-feather in the tail brings with it proportionably increased productiveness; and that, on the other hand, absolute perfection of hen-tailed character in the male bird as generally entails sterility."[4]

The tail is entirely wanting in Rumpless fowls, and it is said that they "are sadly prone to lay unfertilized eggs."[5]

The law of correlation, in its relations to structure

---

[1] "Sheep," p. 141.
[2] *Journal of the Royal Agricultural Society*, vol. xv., p. 285.
[3] "Animals and Plants under Domestication," vol. ii., p. 68.
[4] Tegetmeier on "Poultry," p. 245.
[5] Ibid., p. 232.

and function, furnishes the best explanation of the
difficulty experienced by breeders in retaining and
developing, in their greatest perfection, two essentially different functions in the dominant characteristics
of the same animal.

In attempting to secure the highest development
of some particular quality, a gradual and, it may be,
an undesirable change is so often observed in the
qualities depending on the functional activity of some
other part or parts of the system as to lead to the belief that the quality that is retained is incompatible
with a high development of the function that is impaired in its activity.

A deficiency in the production of milk has often
been noticed in animals that are remarkable in the
tendency to fatten. Mr. Price, a noted breeder of
Hereford cattle, says: "Experience has taught me
that no animals possessing form, and other requisites
giving them a great disposition to fatten, are calculated to give much milk; nor is it reasonable to suppose they should—it would be in direct opposition to
the law of Nature. Had I willed it twenty years ago,
my belief is that I could, by this time, have bred
twenty cows, purely from my own herd, which should
have given a sufficient quantity of milk for (paying)
dairy purposes; and I am equally confident that, in
the same period, I could have bred a similar number
that would not, at any time, have given twenty quarts
of milk per day among them.

"I feel confident I could effect either of these
objects much more easily and certainly than I could
blend the two properties in the same animal, retain-

ing also the form and quality best adapted to live hard and feed." [1]

It is not claimed that high feeding qualities cannot be combined with good milking properties, but that it is easier to excel in either single quality than to secure a high development of both. It does not, as a matter of course, follow that antagonistic characters are strictly incompatible. Additional illustrations of the law of correlation may be found in the chapters relating to other topics; the facts already cited will, however, serve my present purpose, as they clearly indicate that an intimate relation exists between the characters that are comprised in the dominant features of the organization, and that these characters are transmitted in their integrity, without essential change.

An equilibrium of the organization can only be obtained by an arrangement of its elements in strict accordance with the law of correlation. Any modification of even a single character may, therefore, involve corresponding changes in other parts of the system, and a consequent rearrangement of the dominant characteristics.

When the balance of the system is in this manner disturbed, it is difficult to determine the extent of the change that may follow, as it may result in transposing the latent and dominant characters, and develop in the offspring a resemblance to some remote ancestor.

[1] *Farmer's Magazine*, vol. xiv., p. 50. *See* also Culley on "Live-Stock," fourth edition, 1807, p. 87.

CHAPTER VII.

Our domestic animals, in common with other species, are endowed with a flexibility or plasticity of the organization that enables them to adapt themselves to the conditions in which they are placed. As a result of a favorable change in the conditions to which animals are subjected, important modifications of the system are obtained, that we recognize as improvements in form and quality; while deterioration and loss of valuable characters follow when the prevailing conditions of life are unfavorable to the full and healthy development of the organization.

From the fact that variations are more readily produced in domesticated varieties than in wild species, it would appear that the change of conditions involved in the process of domestication has not only produced a wide range of variations in the characteristics of animals, but developed an increased plasticity of the organization that renders them more susceptible to the influence of modifying causes.

The distinguishing characteristics of the various breeds of animals have been produced, in the main, by the modifying influences that prevail in the localities in which they have originated.

In the improved families of pure-bred animals, the influence of artificial conditions in modifying characters is further shown in the excessive development obtained in special directions.

The principal causes of animal variation are climate, food, and habit; the influence of the first two, in many cases, being so intimately connected that it is difficult to determine what is due to each, while all of them may at times act together. Of the many illustrations of the modifying influence of climate that might be drawn from the vegetable kingdom, we shall only present some general statements in regard to two of our leading crops.

Indian-corn (maize) has a wide geographical range, but in its distribution and development it is influenced in a great degree by climatic conditions. In North America its extreme limits at the North " are defined by the isothermal of 67° for July, and it may go beyond 65° for the summer; one month, however, being required at a higher mean than this." [1]

In Northern Europe, including Great Britain, the comparatively low summer temperature prevents the ripening of this valuable cereal, although it is grown in some localities as a forage-crop.

The time required for ripening the crop in localities where it is grown varies greatly with the climate. In its extreme northern range, where the smaller varieties only are grown, but from two to two and a half months are required to bring it to maturity, while at the South a period of from five to six months is necessary.

[1] Blodgett's "Climatology of the United States," p. 420.

Heller, in describing the variations in maize cultivated in Mexico, states that the time of ripening varies
from " seven months to six weeks." [1]

At the North the plant presents a dwarfed appearance, while at the South the stalks are very
large, the ears frequently being higher than a man
can reach.

A collection of corn that I made in 1876, to illustrate the variations produced by climate, represents
many interesting features in the character and distribution of varieties.

At the North the cob, as a rule, is larger in proportion than in the Southern varieties, or in apparently the same varieties grown in the Middle States. At
the North the flint varieties are exclusively grown,
while at the South they are entirely replaced by the
dent varieties. The smallest well-developed ear in
the collection weighs but half an ounce, while the
largest ear turns the scale at one pound eight and a
half ounces.

The influence of climate upon the distribution and
development of wheat is hardly less marked. Samples
in my cabinet from British Columbia, Oregon, Canada,
Michigan, Russia, Norway, Sweden, and Australia,
present marked contrasts in their general appearance.

In North America a mean temperature of from
57° to 65°, and in England of 60°, for the months
of July and August, is required for its full development.

In 1853 the mean temperature of these months in

1 " Patent-Office Report," " Agriculture," 1847, p. 412.

England was from 57° to 59°, which had the effect to diminish the crop from one-half to one-third.[1]

Even peculiarities resulting from a slight difference in locality may have an important influence on the time required for its growth and ripening. Marshall states that, in the Cotswold Hills, a " stone might be flung from the country which sows its wheat in August into that which sows its wheat in December."[2]

A variety of food is required by animals, so that each organ concerned in the process of nutrition may perform its fair proportion of work, and thus secure a healthy development, resulting in a symmetrical balance of the system.

Among animals we cannot fail to observe that the small breeds of sheep and cattle in mountainous regions present a decided contrast to the breeds obtaining an abundant supply of food in the fertile valleys of the same country.[3]

As the relation of the size of animals to the supply of food they are provided with has been noticed by almost every writer on the management of livestock, we need not, for the present, give a detailed discussion of the subject.[4]

---

[1] Blodgett's " Climatology of the United States," p. 446. *See* also *Journal of the Royal Agricultural Society*, 1873, p. 379.

[2] " Rural Economy of Gloucestershire," 1789, vol. ii., p. 52.

[3] Low's " Domestic Animals," pp. 41, 264.

[4] " Agricultural Report of Staffordshire," p. 174; " Agricultural Report of Middlesex," p. 406; Youatt on " Cattle," p. 525; Youatt on " The Horse," p. 60; Coventry on " Agriculture," p. 182; Dickson's " Practical Agriculture," vol. ii., pp. 638–640; Cline on " Breeding and Form," p. 12.

The great development in fattening quality and in early maturity, that characterizes the modern meat-producing breeds of cattle and sheep, has been secured by a liberal supply of nutritious food during the period of growth, in connection with a judicious system of breeding, that has fixed and made dominant the desirable modifications thus obtained.

The Spanish merino sheep, imported into this country in the early part of the present century, were valued principally for their wool, the peculiar system of management to which they had been subjected for many generations having made them decidedly deficient in ability to fatten and in the quality of their flesh.

Their descendants, from the influence of modified habits and a better supply of food, present such a wide departure from the original type, in the greater weight and quality of fleece, in the increased tendency to fatten, and the marked improvement in the quality of flesh, that they are justly entitled to the distinctive appellation of American merinoes which is now generally given them.

The breeders of merino sheep have been directing their attention almost exclusively to the improvement of the fleece, and the greater value of the improved breed for the purposes of the feeder and the butcher has been obtained through the means adopted for the development of other characters.

It is perhaps impossible to obtain any decided modification of a single character without producing corresponding modifications of other parts of the organization.

In the improvement of the mutton-breeds of sheep, breeders have almost uniformly aimed to secure greater symmetry in their general proportions, in connection with early maturity, and to diminish any tendency to coarseness that may have existed in the original breed.

In all of the improved breeds of sheep a general refinement of the system has been developed, as the result of the improvements that have been made in special characters, and this has apparently produced a finer fibre of wool, notwithstanding the lack of attention to this particular quality on the part of breeders.

In 1835 Mr. Youatt, assisted by Mr. Powell, a manufacturer of microscopes in London, made measurements of the wool-fibres of different breeds, which were published in 1840, in his work on "Sheep,"[1] as follows:

	No. of Fibres to the Inch.
Saxon	840
Merino	750 (from Lord Western's flock)
Odessa wool . . .	750
Negretti	750
Common merino . .	750
Australian wool. . .	750
New South Wales wool .	750
McArthur's Australian wool,	780 (Saxon?)
Leicester . . .	500
" (from Ireland) .	560
Cheviot	500
South-Down . . .	660

The finest sample measured was from the Deccan

---

[1] Youatt on "Sheep," p. 87.

black sheep of India, which gave 1,000 fibres to the inch.

In 1864–'65 I measured wool from several flocks, with the following result.

With the exception of the Saxon, of which the date of shearing was not known, the samples were all from fleeces of 1864.

	No. of Fibres to the Inch.	
Saxon ram . . .	1242	(from flock of W. H. Ladd, Ohio).
" ewe . .	1347	" " " " " "
Silesian ram . . .	1352	" " W. Chamberlain, New York.
Merino ram (Silver-mine)	1212	" " E. Hammond, Vt.
" " (Sweepstakes)	1186	" " " " "
" " (Gold-drop)	1185	" " " " "
" ewe (Old Queen)	1275	" " " " "
" " (Queen, 2d)	1183	" " " " "
" " (Queen, 3d)	1138	" " " " "
" " (Queen, 4th)	1223	" " " " "
" " (Queen, 5th)	1274	" " " " "
" ram . . .	1164	" " Hon. Chas. Rich, Lapeer, Michigan.
Merino ewe . .	1064	" " " " "
" " . . .	1164	" " " " "
" " . .	1023	" " " " "
" " . . .	1022	" " " " "
Grade merino ewe .	1077	" " Mich. Agricul. Col.
" " " .	1249	" " " " "
" " " .	1248	" " " " "
South-Down ewe . .	732	" " " " "
" " " .	708	" " " " "
" " " . .	742	" " " " "
" " " .	845	" " " " "

In 1877 I measured samples of wool obtained at

the Centennial Exposition at Philadelphia, in 1876, as follows:

No. of Fibres to
the Inch.

Cheviot owe,	842	from Ed. Henty, Portland, Victoria, Aust.
" "	579	" " " " " "
" lamb	827	" " " " " "
" "	732	" " " " " "
Leicester . .	685	" Wm. Murray, Brie-Brie, " "
" . .	682	" " " " " "
Lincoln . .	769	" Wm. H. Bullivant, " "
" . .	731	" " " " " "
" . .	734	" Ovidio Zubiaurre, Buenos Ayres, Argentine Republic.
Grade Lincoln	874	" Ovidio Zubiaurre, Buenos Ayres, Argentine Republic.
" "	790	" M. Morgan, Argentine Republic.
Merino . .	1199	" R. Goldsbrough, Melbourne, Victoria, Australia.
" . .	1230	" R. Goldsbrough, Melbourne, Victoria, Australia.
" . .	1173	" Thos. Cummings, Victoria, Australia.
" . .	1500	" William Lewis, " "
" . .	1376	" " " " " "
" . .	1079	" Ross and Jas. Randen " "
" . .	1266	" William Lang, Wargam, New South Wales, Australia.
" . .	1325	" Buenos Ayres, Argentine Republic.
" . .	1180	" Wilfren Latham, Los Alamos, Argentine Republic.
" . .	1334	" Wilfren Latham, Los Alamos, Argentine Republic.
" . .	1184	" Mariano Unsué, Buenos Ayres, Argentine Republic.
" . .	1208	" Geronimo Iraizos, Loberia, Argentine Republic.

No. of Fibres to
the Inch.

Merino . . 1450 from J. W. Corrales, Buenos Ayres, Argentine Republic.

Negretti . . 1138 " George Stegman, Buenos Ayres, Argentine Republic.

" . . 1081 " Charles J. Guerrero, Buenos Ayres, Argentine Republic.

" . . 1162 " Francisco Chas, Buenos Ayres, Argentine Republic.

" . . 1266 " Samuel B. Hale, Buenos Ayres, Argentine Republic.

Rambouillet, 1035 " M. Morgan, Buenos Ayres, Argentine Republic.

" 1062 " Emilio Duportal, Buenos Ayres, Argentine Republic.

" 1150 " Emilio Duportal, Buenos Ayres, Argentine Republic.

As these samples, from widely different localities, are, without exception, much finer than the specimens measured by Mr. Youatt, we may safely attribute the change to the same causes that have produced the modifications of form and feeding qualities that characterize all of the improved breeds.

The Kerry cattle of Ireland are a small and hardy race. The scanty supply of coarse food obtained on their native hills, by industrious efforts, gives a slow growth and a late development of the organization, so that the heifers, it is said, do not breed until six or seven years old.

Animals of this breed raised in Massachusetts, under more favorable conditions for development, are larger than the original type, and mature earlier, the heifers breeding at the age of three years.

As the climate of Massachusetts is not so mild and uniform as that of Ireland, we must attribute the changes observed in these cattle to the influence of shelter during the winter, in connection with a better supply of food.

It is a well-known law of the organization that the highest development of any particular organ, or set of organs, can only be attained by their repeated and systematic exercise.

The athlete, as well as the horse in training for a race, must perform an amount of work that taxes the system severely, to secure that strength and development of the muscular system that fit him for the best exhibition of his powers. The highest mental development can only be obtained by severe intellectual effort.

Mr. Darwin has shown that the proportional weight of the wing-bones of wild-ducks is greater than in tame varieties, while the proportional weight of the leg-bones is greatest in the latter.[1]

The activity of the glandular system depends largely upon the demands made upon it, in accordance with the same principle.

Dr. Carpenter, in his article on the " Varieties of Mankind," says : " Another remarkable fact, relative to the oxen of South America, is recorded by M. Roulin. In Colombia the practice of milking cows was laid aside, owing to the great extent of the farms and other circumstances. In a few generations the natural structure of the parts and the natural state of the function have been restored, the secretion of milk

[1] " Animals and Plants under Domestication," vol. i., p. 345.

taking place only so long as the calf remains with
the mother, and ceasing if it dies or is removed.
Hence we have a valuable confirmation of the belief previously entertained, that the continued production of milk by the European breeds of cows is
a modified function in the animal economy, origi-
nating in an artificial habit kept up through many
generations, and dependent upon a modification of
structure which that habit has been the means of in-
ducing."[1]

The practice, too generally prevailing, of raising
young animals by means of nurses, so that the mothers
may go "dry" and be fitted for exhibition, must result, in a few generations, in a serious deficiency of
the milking qualities.

Sir Charles Lyell informs us that "some Englishmen engaged in conducting the mining operations of
the Real del Monte Company, in Mexico, carried out
with them some greyhounds of the best breed, to hunt
the hares which abound in that country. The great
platform which is the scene of sport is at an elevation of about nine thousand feet above the level of
the sea, and the mercury in the barometer stands ha-
bitually at the height of about nineteen inches. It
was found that the greyhounds could not support the
fatigues of a long chase in this attenuated atmosphere, and before they could come up with their prey
they lay down gasping for breath; but these same animals have produced whelps which have grown up and
are not in the least degree incommoded by the want
of density in the air, but run down the hares with as

[1] "Cyclopædia of Anatomy and Physiology," vol. iv., p. 1312.

much case as the fleetest of their race in this country." [1]

In the modifications of form, habits, instincts, and
general activity of the functions of organs, resulting
from the agencies under consideration, the principle
of correlation, to which we have already referred, may
be readily traced.

We cannot, in fact, make a decided change in any
part of the system without producing a corresponding
modification of some other part that is correlated
with it.

The tendency to early maturity, which is so highly
developed in the meat-producing breeds, is accompanied with a change in the period of dentition, and
this fact has to be taken into account in determining
the age of animals by the teeth. [2]

There is not only a difficulty in producing a considerable modification of several characters at the same
time, but there is also the danger of suppressing some
character we wish to retain, by the development of a
new one not in harmony with it.

Family characteristics are produced by limiting
the range of variations to the particular standard the
breeder wishes to establish. The greatest skill will
be required in establishing the family type, to retain,
in connection with the desired characters, the qualities that give vigor to the constitution and insure
an active performance of the function of reproduction, and to prevent, at the same time, the develop-

---

[1] Quoted from "Cyclopædia of Anatomy and Physiology," vol. iv.,
p. 1303.

[2] *Journal of the Royal Agricultural Society,* vol. xv., p. 323.

ment of peculiarities that are in themselves objectionable.

From the manner in which family characters are produced, it will be exceedingly difficult to ingraft any new character upon a family without destroying, to a greater or less extent, its specific characteristics. In the improved breeds, and especially in those in which early maturity and the tendency to lay on fat are highly developed by artificial treatment, the great predominance of one group of characters seems to involve an unstable condition of the organization, and a consequent tendency to further variation.

It is often remarked that it is more difficult to retain a given character than to produce it. If the conditions that gave rise to a particular character are changed, the character itself must be changed also. It is a common mistake of those not familiar with the principles of breeding and the causes of variation, to suppose that the highly-artificial characters of improved breeds can be retained in the absence of the conditions that produced them.

If high feeding has developed a variation in a particular direction, a scanty supply of food would certainly destroy it, and produce a variation of an opposite character. Improved characters can only be made permanent by breeding together the animals that possess them, and continuing without variation the same system of management that originally produced them.

Improvements that have been effected by better care and an abundant supply of food for many generations, may be lost in a comparatively short time, by placing the animals under less favorable conditions and diminishing their supply of food. A single illustration of the effects of neglect will be given :

" During the French Revolutionary War the excessive price of corn attracted the attention of the Glamorganshire farmers to the increased cultivation of it, and a great proportion of the best pastures were turned over by the plough—cattle were almost entirely neglected. . . . The natural consequence of inattention and starvation was, that the breed greatly degenerated in its disposition to fatten, and, certainly, with many exceptions, but yet as their general character, the Glamorganshire cattle became and are flat-sided, sharp in the hip-joints and shoulders, high in the rump, too long on the legs, with thick skins, and a delicate constitution."[1]

" It is well known that defective sanitary arrangements in the dwellings of the poor may, by primarily affecting the parents, impair the physical development of their offspring, and that congenital deformities are, for example, sometimes the result of the continued deprivation of light, which thus indirectly induces an arrest of development, such as can be produced directly and at will in the case of tadpoles, which, in the absence of light, fail to become frogs."[2]

" The effect of darkness in producing deformities is well illustrated in the case of the French historical painter, Ducornet, who used to paint with his feet, having been born without arms, of poor parents living in one of the dark caverns under the fortifications of

[1] Youatt on "Cattle," p. 51.
[2] Sedgwick, in *British and Foreign Medico-Chirurgical Review*, July, 1863, p. 174.

Lille. It appears that several of the deformed beggars in Paris had also been born at Lille, and that the effect of the absence of light in these underground places, in producing malformed births, was so notorious that the magistrates of Lille issued strict orders to prohibit the poor from taking up their abode in them."[1]

Variations frequently occur in particular localities, that cannot be explained on account of the obscure action of the agencies that produce them. Such variations are said to be the result of endemic influences, which is a convenient name for local agencies that are not as yet fully understood.

As an illustration of the obscure action of endemic causes of variation, the following examples are given : In the case of a family which dwelt alternately at Paris and Bordeaux, "the children engendered at Bordeaux were all born deaf-mutes; the children engendered at Paris were all endowed, as their parents, with perfect integrity of hearing. And this endemic influence is still more clearly shown in the case recorded by Puybonnieux ('Mutisme et Surdité,' p. 30, 1846), of a married couple with eight children, of whom five were deaf-mutes; four of these last and two children who could speak were born at Rebrechien, at a house called *Le Jeu de Paume*, situated near the forest of Orleans, in a place elevated and apparently healthy; nevertheless, the people who had dwelt there before the married couple referred to, had had three children, of whom two were deaf-mutes."[2]

[1] *Medical Gazette,* vol. x., p. 818, 1832 ; quoted by Sedgwick in footnote, *loc. cit.,* p. 174.

[2] *British and Foreign Medico-Chirurgical Review,* July, 1863, p. 175.

The development of special characters in our domestic animals, and their consequent improvement in a particular direction, is apparently limited by the tendency to diverse variations, from the increased sensitiveness of the organization to the influence of modifying agencies, and the defective equilibrium of the organization arising from the excessive predominance of a single character. If a variation in a special direction is made at the expense of constitutional vigor, integrity of the nutritive organs, and fecundity, it becomes an abnormal character that cannot be perpetuated.

CHAPTER VIII.

THE conditions of the animal organization that have an influence upon the function of reproduction seem to require more than a passing notice. The fertility of animals is frequently influenced by changes in their surroundings and habits that cannot, in themselves, be considered unfavorable to the healthy action of the system.

It has been observed that the procreative powers are impaired, or even entirely wanting, in many wild species, when placed in confinement. The elephant, the tiger, squirrels, monkeys, parrots, and many other animals, it is said, rarely, if ever, breed when subjected to man's control. Mr. Darwin, on the authority of Mr. Bartlett, records the remarkable fact that "lions breed more freely in traveling collections than in the Zoölogical Gardens."[1]

The flying-squirrel, when breeding in captivity, has not been known to produce more than two young at a birth, while in a state of nature it produces from three to six.[2]

"The African ostrich, though perfectly healthy

[1] "Animals and Plants under Domestication," vol. ii., p. 185
[2] Darwin, loc. cit., p. 187.

and living long, in the south of France never lays more than from twelve to fifteen eggs, though in its native country it lays from twenty-five to thirty."[1]

Lord Somerville says the Spanish merino sheep, in England, when first imported, had a tendency to barrenness and " there was a great deficiency of milk in the ewes,"[2] which he attributes to the severe journeys the sheep were accustomed to make in Spain. As a deficiency in the secretion of milk and a tendency to barrenness have not been observed in these sheep when removed to other countries, these defects in England must have been owing to a change in the conditions of life, rather than to a previous habit of the system.

According to M. Roulin, " in the hot valleys of the equatorial Cordilleras, sheep are not fully fecund," and geese, taken to the lofty plateau of Bogota, did not at first breed well.[3]

Mr. Darwin says: " In Europe close confinement has a marked effect on the fertility of the fowl: it has been found in France that, with fowls allowed considerable freedom, only twenty per cent. of the eggs failed; when allowed less freedom forty per cent. failed; and, in close confinement, sixty out of the hundred were not hatched."[4]

Mr. Darwin was assured that " those animals which usually breed freely under confinement, rarely

[1] Darwin, *loc. cit.*, p. 191.

[2] Somerville's " Facts and Observations," p. 14 ; quoted in Youatt on "Sheep," p. 181.

[3] " Animals and Plants under Domestication," vol. ii., p. 197.

[4] *Loc. cit.*, p. 198.

breed, in the Zoölogical Gardens, within a year or two after their first importation," and he adds that, "when an animal which is generally sterile under confinement happens to breed, the young apparently do not inherit this power, for, had this been the case, various quadrupeds and birds, which are valuable for exhibition, would have become common." [1]

"The carnivora in the Zoölogical Gardens were formerly less freely exposed to the air and cold than at present; and this change of treatment, as I was assured by the former superintendant, Mr. Miller, greatly increased their fertility." [2]

From the preceding statements it might be inferred that the state of domestication was not favorable to fertility; but we find, nevertheless, that domesticated varieties are more prolific than wild species. Tame geese and ducks lay many more eggs than wild ones. Dogs have a larger number of young at a birth than their wild cousins, the wolf and the fox.

The tame varieties of swine are more prolific than wild species. "The wild rabbit is said generally to breed four times yearly, and to produce from four to eight young; the tame rabbit breeds six or seven times yearly, and produces from four to eleven young."

Wild pigeons do not breed so often as tame varieties, and Macgillivray states that, while the wild rockpigeon breeds but twice a year, "the same pair, when tamed, generally breed four times." [3]

[1] *Loc. cit.*, p. 195.
[2] Ibid., p. 185.
[3] "Principles of Biology," vol. ii., p. 457; "Animals and Plants under Domestication," vol. ii., p. 139.

The greater fecundity of domesticated varieties, as compared with that of wild species, is, in great measure, owing to a better supply of food throughout the year, and the more uniform conditions in which they are placed.

The activity of the reproductive organs is necessarily dependent upon the function of nutrition which supplies the materials concerned in its operations.

Dr. Carpenter says, "There is a certain degree of antagonism between the nutritive and the generative functions, the one set being executed at the expense of the other."[1]

A certain activity of the nutritive functions is required to secure the greatest fertility in both plants and animals. When the function of nutrition is impaired by disease, or when the supply of food is not sufficient for the wants of the system, the reproductive powers suffer a corresponding decrease in their activity.

Sheep bred on rich pastures are more likely to produce twin lambs than those gaining a scanty subsistence in less favored localities.

It is said that, "among the barren hills of the west of Scotland, two lambs will be borne by about one ewe in twenty, whereas in England something like one ewe in three will bear two lambs."[2]

While full feeding seems to increase the fecundity of varieties, any excess in the nutritive activity of the system will as readily impair the powers of reproduction.

[1] "Comparative Physiology," p. 147.
[2] "Principles of Biology," vol. ii., p. 459.

In flowering plants, "it is well known that an over-supply of nutriment will cause an evolution of leaves at the expense of the flowers, so that what actually would have been flower-buds are converted into leaf-buds; or, the parts of the flower essentially concerned in reproduction, namely, the stamens and pistil, are converted into foliaceous expansions, as in the production of 'double' flowers from 'single' ones by cultivation; or, the fertile florets of the 'disk,' in composite species, such as the dahlia, are converted into the barren but expanded florets of the 'ray.' And the gardener who wishes to render a tree more productive of fruit is obliged to restrain its luxuriance by pruning, or to limit its supply of food by trenching around the roots."

"During the period of rapid growth, when all the energies of the system are concentrated upon the perfection of its individual structure, the reproductive system remains dormant, and is not aroused until the diminished activity of the nutritive functions allows it to be exercised without injury to them."[1]

While the period of rapid growth is not favorable to the development of the reproductive powers, from the great preponderance in the system of the nutritive functions, it will also be found that any marked deficiency in the processes of nutrition, as in the decline of life, will result in a decrease and final loss of fertility. The age of an animal will thus have an impor-

---

[1] Carpenter's "Comparative Physiology," p. 147. Root-pruning, as a remedy for "unfruitfulnesse in trees," was recommended by Sir Hugh Plat, in his "Garden of Eden," fifth edition, published in London, 1659, p. 162.

tant influence on fecundity, through the variations involved in the nutritive functions.

In a preceding chapter (page 36) the relations of age to fecundity have been noticed, in discussing the influence of immaturity in the parents upon the development of their offspring.

It was there shown that the eggs of young animals were comparatively small and few in number.

The sow and the bitch, breeding at an early age, have comparatively few young in a litter; at the period of maturity the number reaches a maximum, and, at an advanced age, the number is diminished. "The young hamster produces only from three to six young ones, while that of a more advanced age produces from eight to sixteen." [1]

Similar variations in the number of young at different ages have been observed in other animals.

The quality of food seems to exercise an influence on the reproductive functions, but the data for a full discussion of the subject are as yet wanting. In the development of the bee, the form of the cell and the character of the food determine the fertility or non-fertility of the perfect insect, and it is also claimed that in insects the sex is, in some cases, determined by the process of nutrition. [2]

A large proportion of sugar in the food is supposed to interfere with the reproductive functions. [3]

Prof. Tanner, in his paper on the reproductive powers of animals, says: "The general system of diet

---

[1] "Principles of Biology," vol. ii., p. 438.
[2] *The Popular Science Monthly*, April, 1874, p. 761.
[3] *Journal of the Royal Agricultural Society*, 1865, p. 267.

must also be looked upon as taking its share in influencing the reproductive functions. When the fall of rain has been small, and the herbage more than usually parched, we find unusual difficulty in getting ordinary farm-stock to breed—a dry dietary is very unfavorable for breeding animals, and very much retards successful impregnation. On the other hand, rich, juicy, and succulent vegetation is very generally favorable to breeding. Apart, therefore, from the direct influence of the food given, it is certain that the condition in which it is consumed materially influences the breeding powers." [1]

Mr. Mills, in his " Treatise on Cattle," published in 1776, remarks that " mares which have been brought up in the stable on dry food, and afterward turned to grass, do not breed at first; some time is required to accustom them to this new aliment." [2]

In the wild species that breed twice a year it has been stated that the time of breeding is determined by the abundance of food; but this does not appear to be the case with migratory birds, in which the impulse to nest-building and migration occur together, at an early period in the spring, before they can obtain an abundant supply of food.

There seems to be a marked relation between the size of animals and their fecundity, which may perhaps be owing, in part at least, to the modifying influence of the nutritive functions. Throughout the entire animal kingdom the small species of animals appear to be more prolific than large ones, and, as a rule,

[1] *Journal of the Royal Agricultural Society*, 1865, p. 269.
[2] *Loc. cit.*, p. 66.

they breed at an earlier age, and at shorter intervals, and produce a greater number of young at a birth.

The elephant, the rhinoceros, the hippopotamus, the camel, and the dromedary, produce but one at a birth; the cow, the red-deer, the sheep, the llama, the mare, and the ass, produce one or occasionally two; the goat, the roe-deer, and the chamois, produce two or three; the cat, the fox, the jackal, the tiger, the lion, and the bear, produce from two to six; the dog and the wolf, from five to ten; the wild-boar, from four to ten; and the domestic sow, from eight to seventeen; while the smaller rodents have produced as many as nineteen young at a birth.

The larger animals, as the great pachyderms, the solipeds, and the ruminants, breed but once a year; while the smaller mammals breed two or three times in a year.[1]

Among mammals, swine, and a few domesticated varieties, present almost the only exceptions to the prevailing inverse relation of size to fecundity. The larger birds are less prolific than the smaller species, while among the most minute members of the animal kingdom the most astonishing fecundity is observed.

In the cases of diminished fecundity from over-feeding, or from an abnormal activity of the nutritive functions, a plethoric condition of the system is produced that may, in itself, impair the vigor of the reproductive powers, or lead to the development of local

[1] Colin, "Physiologie comparée," tome ii., p. 531; Spencer's "Principles of Biology," vol. ii., pp. 435, 436; "Animals and Plants under Domestication," vol. ii., p. 139.

congestion and inflammation that interfere with the activity of the function.

In quite a number of cases of barrenness, in highly-fed and plethoric animals, that have come under my observation, the defect was clearly attributable to an extreme irritability of the organs of generation, resulting from congestion or local inflammation.

In some of these cases congestion of the mucous membrane of the vagina and mouth of the uterus was the only abnormal peculiarity that could be detected, while in others there was congestion of the ovaries, or deposits of tuberculous matter involving a large proportion of their tissues.

In this connection, we should not overlook the fact that the highly-artificial conditions to which animals are subjected, to secure the development of special characters, render the system exceedingly sensitive to the influence of the acknowledged causes of the scrofulous habit.

Even when the unsymmetrical development of the organization does not proceed far enough to produce an unhealthy condition of any of the reproductive organs, it may constitute a predisposing tendency to disease that is liable to be made active by slight exciting causes.

A long series of derangements of the organs of generation, of every grade of intensity, may thus arise, directly or indirectly, through the influence of the defective equilibrium of the system, produced by pampering and over-feeding.

If the procreative functions are impaired by a plethoric condition of the system, without complications from local disease, the defect may be corrected in many instances by active exercise, low diet, or depletion ; but when the local derangements of the system are the result of disease it will be difficult to restore the normal activity of the function, even under the most skillful treatment.

A remarkable development of the tendency to lay on fat is usually accompanied by a delicacy of constitution, a diminished secretion of milk, and a loss of fecundity.

It is a popular notion that very fat animals are not likely to be good breeders, and when, even in flocks and herds that are not highly bred, a marked tendency to lay on fat is observed in precocious females, their ability to breed is often called in question. The general prevalence of such opinions seems to indicate that experience has shown that the excessive production of fat is incompatible with a high development of the reproductive powers ; and it is for this reason that objections are made to what is called "show condition" in breeding stock.

Prof. Tanner, one of the best authorities on this subject, says : "The non-impregnation of the female may generally be traced to an excessive fatness in one or both of the animals, and an absence of constitutional vigor. The breeding powers are most energetic when the animals are in moderate condition, uninfluenced either by extreme fatness or leanness."[1]

The antagonism of the reproductive functions and the "fatty diathesis" is shown in the fact, well known to feeders, that the removal of the ovaries of the fe-

[1] *Journal of the Royal Agricultural Society*, 1865, p. 265.

male, or of the testicles of the male, gives an increased tendency to fatten.

The influence of an excessive deposition of fat in the tissues upon the general health and activity of the system is thus referred to by Dr. Cragie, in his paper on "Adipose Tissue:"

"In persons of this description, who, it is matter of common observation, are generally not only plethoric but bloated, and liable to imperfect circulation, and disorders of the circulation and secretions generally, and in whom very slight causes often induce serious disorders, the adipose tissue appears to lose a great proportion of the small degree of vital energy which it possesses; and the more abundant its secreted product is, the less active are its vessels and the inherent properties of the membrane.

"In consequence of this greatly-impaired energy, slight causes, as cold, injury, punctures, etc., produce suddenly a complete loss of circulation and action in the tissues—for it is not increased but diminished action—and this impaired energy continues until the natural function of the tissue becomes extinct." As to the formation of fat, he adds: "In females and eunuchs it is more abundant than in males; in females deprived of the ovaries it is more abundant than in those possessed of those organs, and it is well known that sterility is frequent among the corpulent of both sexes." [1]

In many instances the integrity of important organs is impaired by deposits of fat, or by the actual transformation of their substance into fatty tissue,

[1] "Cyclopædia of Anatomy and Physiology," vol. i., p. 62.

which is known to medical men as "fatty degeneration." Dr. Carpenter says: "There is one remarkable form of degeneration, however, which is common to nearly all tissues, and which seems to occur, as a normal alteration, in many of them at an advanced period of life; this consists in the conversion of their albuminous or gelatinous materials into fat, thus constituting what is known as fatty degeneration. That this change is not due to the removal of the normal components of the tissues, and the substitution of newly-deposited fatty matter in their place, but is (in most cases at least) the result of a real conversion of the one class of substance into the other, has been already pointed out;" and he further remarks that "there is reason to believe that 'fatty degeneration,' the form under which degeneration most commonly presents itself, is in reality far more frequent than simple wasting of the tissues; but it attracts less notice because their bulk is little or not at all diminished, and it is only when their function becomes impaired that attention is seriously drawn to the change." [1]

Dr. Flint, one of the best authorities on the subject of physiology, says fat "does not take part in the nutrition of the parts that are endowed, to an eminent degree, with the so-called vital functions; and, when these tissues are brought to the highest point of functional development, the fat is entirely removed from their substance. Long disuse of any part will produce such changes in its power of appropriating nitrogenized material for its regeneration that it soon becomes atrophied and altered. Instead of the normal nitro-

[1] "Human Physiology," pp. 553, 559.

genized elements of the tissue, we have, under these circumstances, a deposition of fatty matter. The fat is here inert, and takes the place of the substance that gives to the part its characteristic function. These phenomena are strikingly apparent in muscles that have been long disused or paralyzed, or in nerves that have lost their functional activity. If the change be not too extensive the fat may be made to disappear, and the part will return to its normal constitution by appropriate exercise ; but frequently the alteration has proceeded so far as to be irremediable and permanent." [1]

The reproductive organs of very fat animals are frequently affected with fatty degeneration, to an extent that impairs or entirely destroys their functional activity.

In a valuable paper on " The Reproductive Powers of Animals," Prof. Tanner says : " For the purpose of more fully investigating the causes of barrenness, I have examined the ovaries of several heifers which were, after a very careful trial, condemned and killed as barreners, and I have every reason to believe that by far the larger proportion were naturally quite competent for breeding, and that, in the majority of cases, non-impregnation arose from the seminal fluid never reaching the ovum, which was ready for fertilization, or from that fluid not being of a healthy character.

" In some cases in which the ova were, to all appearances, perfectly healthy, the tubes—whereby the seminal fluid should have been conveyed—were so

[1] Flint's "Physiology of Man"—"Nutrition," p. 381.

overcharged with fatty matter that impregnation was rendered impossible.

" In other cases the ovaries were in an unhealthy condition, either one or both having, to a great extent, wasted away. Sometimes one of the ovaries had been suffering from atrophy, and the other in such an irritable and sensitive condition that it might be almost described as inflamed, and under such circumstances the formation of a healthy ovum could be scarcely expected. In other instances the ovaries had become considerably enlarged, in consequence of a *fatty degeneration* of these organs having taken place." [1]

It is to be regretted that the condition of these animals, in regard to fattening tendency and constitutional peculiarities, is not given in the above cases, as it would aid us in determining the cause of the observed pathological conditions. Of the cases of barren females that I have had an opportunity to investigate, the defect was attributable, in about equal proportions, to fatty degeneration of the ovaries, scrofulous tumors of the ovaries, and congestion and chronic inflammation of the uterus and its appendages—all of which were apparently the result of an excessive development of the tendency to fatten.

When the fatty degeneration, or the scrofulous tumors, were confined to one ovary, its fellow was usually the seat of congestion or chronic inflammation, and thus unfitted to develop a healthy ovum.

From the correlated relations of the functions of nutrition and reproduction, it will be seen that great activity of the fat-producing functions, even when not accompanied by local disease, will involve a corresponding decrease in the activity of the reproductive powers.

From the antagonistic conditions presented in the law of correlation, it will perhaps be impossible to secure the highest type of perfection in the production of fat without impairing, to some extent at least, the functions of the reproductive organs.

We have already noticed the apparent incompatibility of the fat and the milk producing functions, and we find also that a diminished secretion of milk is often observed in animals that are not prolific, while the best breeders are usually good milkers. Prof. Tanner, in his paper which we have already noticed, says: "The formation of milk is intimately connected with the reproductive powers. The secretion of milk is dependent upon the activity of the mammary glands, and these are either under the direct influence of the breeding-organs, or else they sympathize very closely with them. Those animals which breed with the least difficulty yield the best supplies of milk, and produce the most healthy and vigorous offspring.

"Now, it must be admitted that, however much we have improved the symmetry and feeding power of stock, we have suffered them to deteriorate in value as breeding animals, by the decrease of their milking capabilities. In proportion as we adopt a more natural system of management, for the purpose of keeping stock in a healthy and vigorous breeding condition, so shall we reap the indirect benefit of a better supply of milk. It is true that a deficiency in the yield of milk may be met by other resources, but, since a short sup-

ply of milk is indicative of, and associated with, en-
feebled breeding powers, every care should be taken
to obviate this defect."[1]

In the human family, as the physical organization,
in structure and function, does not essentially differ
from that of the lower animals, the same causes of
impaired fertility will be operative, if the habits and
conditions of life do not present a wide departure
from those that prevail in a state of nature.

With an advance in civilization, however, when
the mental faculties attain a high degree of development, and the physical activity of the system is intensified through the action of the nervous system, a new
element of variation is introduced, that disturbs the
equilibrium of the system and increases the activity
of the various causes that interfere with the procreative functions.

In a work published nearly one hundred years ago,
Dr. Black remarked that "high refinement is an ob-
stacle to propagation." In a paper read before the
Statistical Society, in 1843, Sir John Boileau says:
"It is a fact that rich families, taken in general, are
those which have the fewest children; and their ranks
would become thinner, generation after generation, if
they were not gradually recruited by new families of
recently-acquired wealth.

"The effect which riches have in restraining the
fecundity of marriages is nowhere more apparent than
in Paris. The most opulent families of France con-
gregate there, and, as they select certain quarters of
the town for their residence, the facts brought out in

[1] *Journal of the Royal Agricultural Society*, 1865, p. 270.

them are more remarkable and complete than any-
where.

"Now, by the investigations made under the direction of the Comte de Chabrol, the average of births
to a marriage is, in the different arrondissements, in
regular inverse proportion to the easy or opulent circumstances of the population. In the first four arrondissements united, which are those where the most
opulent families reside, the number of children to a
marriage is only 1.97; that of the four poorest arrondissements, on the contrary, is 2.86; and the difference between the two arrondissements placed at the
extremities of the scale is as 1.87 to 3.23, or more
than 73 per cent.

"These facts deserve the more attention because, in
spite of the reasons which determine the inhabitants
of Paris to choose peculiar localities, according to their
respective circumstances, some poor families will be
found in the quarters inhabited by the rich, and some
rich families in the quarters occupied by the poor;
which fact necessarily diminishes the difference we
should establish if it were possible to separate completely the different classes of the population. We
arrive at this important consideration that, if the
second, third, tenth, and first arrondissements, where
the richest families in Paris reside, were not continually recruited from families freshly acquiring wealth,
the actual number of inhabitants would not be maintained. Not only the children born there are less
numerous than their parents, but, as we must deduct
those who die in infancy, or who never marry—and
that we must estimate these at least at a quarter of

the whole, in a town where thirteen children out of twenty-nine do not live to twenty-one — it follows that in three generations, or the space of a century, the population would be reduced to half its number."

Mr. G. R. Porter, in his "Progress of the Nation," says, "Frequently, and indeed almost always, in old-settled countries, the proportionate number of births decreases with the advance of civilization and the more general diffusion of the conveniences and luxuries of life."[1]

Other writers speak of the generally acknowledged influence of the plethoric condition of the system that prevails among the wealthy, in producing diminished fecundity. It does not, however, follow from the facts stated that privation and want are favorable to fertility, as the reverse is true. It is a well-known fact that famines not only diminish population by an increased death-rate, but also by a diminution of the birth-rate.[2]

In the absence of those special conditions that antagonize the procreative functions, the greatest fecundity may be expected when the food-supply is sufficient for the wants of the system, and active habits of life conduce to a healthy performance of the various organic functions.

There are facts that seem to show that an improved condition of the system, resulting from a better food-supply after a period of privation and even of disease,

---

[1] The last three quotations have been copied from Walford's "Insurance Cyclopædia," vol. iii., pp. 185, 190.

[2] Walford, "Insurance Cyclopædia," vol. iii., p. 163.

may produce an unusual activity of the functions of reproduction.

The unusually rapid increase of population, after a country has been scourged by a famine or pestilence, has often been remarked.

After the plague of 1348 in England, the "flocks and herds wandered about at will, without herdsmen, shepherd, or owner," and labor was so scarce that landlords were glad to have their lands cultivated by their tenants without payment of rent. Population, however, speedily righted itself. "We are told that after the plague double and triple births were frequent, that most marriages were fertile, and that no serious effects were produced, in a short time, on the numbers of the people."[1]

In examining the various causes of impaired fecundity, we must not lose sight of the influence of the transmission of ancestral tendencies and peculiarities.

If the ancestors of an animal are not prolific, it will inherit a bias of the organization that is favorable to the action of the various causes of sterility and barrenness; that is, the natural tendency or predisposition of the organization will, as it were, add to the intensity of the forces that interfere with the normal performance of the function of reproduction, and thus aid in its suppression.

The production of twins will be found to depend, not only upon the supply of food, as already noticed, but on peculiarities of the system that have been inherited.

---

[1] Rogers, "History of Agriculture and Prices," vol. i., pp. 299–301.

" Osiander [1] relates the case of a woman who, in eleven *accouchements,* had given birth to thirty-two children, was herself born with three other twins, and her mother had had thirty-eight children; another woman, delivered of five children at a birth, had a sister who was delivered of three; and lately at Rouen, twin sisters gave birth to twins on the same day.[2] Mr. J. Lewis Brittain [3] related last year, at the Edinburgh Obstetrical Society, the case of a woman who had twins eleven times, and whose mother had had twins twice; and the report states that ' several of the members mentioned that they knew of some analogous cases.' "

" Dr. Mitchell, in a paper on ' Plural Births in Connection with Idiocy,'[4] cites the following cases: The mother of an idiot, twin-born, bore twins twice, the maternal grandmother once, one maternal aunt twice, another once, and a sister once; in a second case the mother was herself one of twins, and she bore twins once, and, in a third case of a twin-born idiot, the aunt had borne twins; while among the cases in which the idiot was not twin-born, in one the mother and the maternal grandmother each bore twins twice; in a second case the mother and the maternal grandmother each bore twins once, and a maternal aunt twice; in a third case the mother and three maternal aunts each bore twins once; in a fourth case the mother bore twins once, and a maternal aunt bore

---

[1] "Handbuch der Entbindungskunst," Band i., pp. 316, 317.
[2] *British Medical Journal,* November 30, 1861, p. 598.
[3] *Edinburgh Medical Journal,* 1862, p. 468.
[4] *Medical Times and Gazette,* 1862, p. 513.

twins four times running; and in a fifth case the
mother and two sisters of the idiot each bore twins
once.

"It is also well known that the hereditary production of twins in sheep is encouraged by saving the
ewe-lambs that are twins.' Notwithstanding these
facts, there are some cases which show that twins occasionally owe their descent as such to the male line,
of which the following case affords a good illustration : Two brothers (twins) both had twins by their
wives many times in succession; the wife of one of
them having died, the second wife produced, like the
first, twins;' and, in the case recorded by Mr. Stocks,'
of Salford, twin brothers also produced twins ; one of
them having a family of ten children, eight daughters
and two sons, all of whom were twin-born; and the
other a family of eleven children, of whom eight were
twin born ; it is, moreover, to be noticed that in this
last case, while five of the female twins in the succeeding generation produced twins at their first birth,
the three children of the only one of the male twins
of whom any account is given, were all born singly,
leading us to infer that the hereditary predisposition
to twins was probably derived from a female ancestor, and that each of the twin brothers referred to, in
addition to being the medium of transmission, also
shared in the inheritance.

"In connection also with the influence of sex in the

---

¹ "Notes on Fields and Cattle," by Rev. W. H. Beever, 1862, p.
144.

² "Nouvelle Dictionnaire d'Histoire Naturelle," tome xii., p. 566.

³ *Lancet*, July 20, 1861, p. 78.

production of twins, it is necessary to notice the popular error respecting the alleged barrenness of females who have themselves been born as twins with male children, for it is still customary among nurses and midwives, in some places, to talk somewhat disrespectfully of such females, as disqualified for the marriage state, in consequence of their supposed inability to have children.

"This error, which probably arose from the well-established fact of the barrenness of the free-martin (the imperfect cow-calf twin with a bull-calf), was refuted by Mr. Cribb, in a paper published in 1823,[1] which contains six cases of such females becoming mothers.

"Dr. Sieveking has informed me of a case in which a woman, twin with a male, subsequently gave birth to twins; and any remaining doubt on the subject is removed by the fact that such females have on some occasions become even more than usually prolific, as in the case which occurred near Maidenhead,[2] of quadruplets, consisting of three boys and one girl, who were all reared, and the only female in this quartet subsequently became the mother of triplets, consisting of two boys and one girl."[3]

The following remarkable case would need to be well authenticated to entitle it to credence: "The

---

[1] London *Medical Repository*, 1823, pp. 213–216.

[2] Ibid., 1827, p. 350.

[3] In the above quotation from Mr. Sedgwick's paper, *British and Foreign Medico-Chirurgical Review*, July, 1863, pp. 170, 171, the original sources from which the cases were compiled are cited in the footnotes.

Boston Medical and Surgical Journal stated that on the 21st of August, 1872, Mrs. Timothy Bradler, of Trumbull County, Ohio, gave birth to eight children —three boys and five girls. They were all living and healthy, but *quite small.* She was married six years previously, and weighed two hundred and seventy-three pounds on the day of her marriage. She has given birth to two pairs of twins, and now eight more, making twelve children in six years. Mrs. Bradler was one of a triplet, her mother and her father being twins, and her grandmother the mother of five pairs of twins." [1]

" In a remarkable instance which occurred in the city of New York, the mother had twelve children within four years after her second marriage, at four births, there having been twins at the first, triplets at the second and third, and quadruplets at the fourth. The first (twin) birth occurred at the age of thirty-five; she had previously given birth to seven children, one only at a time." [2]

" A still more remarkable case occurred in Mercer County, Pennsylvania, in 1816, ten children having been born within twelve months, five at each of two births. The mother died about a year after the second birth, but meantime gave birth to twins; or twelve children in twenty months. She was thirty-seven years old at her death." [3]

---

[1] *British Medical Journal*, November, 1872, as quoted in Walford's " Insurance Cyclopædia," vol. iii., article " Fecundity," p. 200, where a large number of cases of multiple births are recorded, including twenty-five cases of triplets, thirteen of quartets, three of quintets, and one each of six, eight, and ten, at a birth.

[2] Dr. E. R. Peaslee, Johnson's " Universal Cyclopædia," article "Gestation."   [3] Ibid., *loc. cit.*

" An instance is mentioned in the *Bulletin des Sciences* of a cow belonging to a French agriculturist, which produced nine calves at three successive births, namely, four at the first, three at the second, and two at the third; all of which, except two of the first birth, grew up and were nursed by the mother; but the heifers afterward produced each only a single calf." [1]

According to Culley, the Teeswater ewes "generally bring two lambs each, and sometimes three; there are instances of even four or five, as was the case with Mr. Edward Eddison's ewe, which, when two years old, in 1772, brought him four lambs; in 1773, five; in 1774, two; in 1775, five; in 1776, two; and in 1777, two. The first nine lambs were lambed within eleven months." [2]

A ewe belonging to James Wilkie, Esq., of the county of Berwick, Scotland, "produced eleven lambs in the course of three immediately succeeding seasons. In the spring of 1803, she had four lambs; in 1804, three; and again four in 1806. She was of the ordinary breed of the lower part of the country." [3]

Mr. Kerr remarks in regard to these sheep that "ill-fed ewes hardly ever have twins, while those that are in good condition, when put to the ram, very often have twins, and sometimes triplets."

" In 1806, in a flock of Norfolk ewes belonging to Mr. Wythe, of Eye, one on the 18th of February yeaned three lambs; on the 20th another dropped three, and a third, five, on the 21st; a fourth, four, on the 23d;

[1] " British Husbandry," vol. ii., p. 438, note.
[2] Culley on " Live-Stock," p. 123.
[3] Kerr's " Agricultural Survey of Berwickshire," p. 403.

and on the same day a fifth produced three. On the 25th a sixth ewe dropped three lambs; and a seventh yielded four on the 27th. So that seven ewes yeaned twenty-five lambs, all of which were reared." [1]

"Last week a ewe, belonging to Mr. Kitter, yeaned five lambs; she also brought five lambs last year, and four the year before; i. e., fourteen lambs in three years, and not a weak or deformed one in the whole number." [2]

"Mr. Meadows, of Salcey Forest, Northampton-shire, has a ewe which brought him three lambs in 1802, four in 1803, four in 1804, and four in 1805; being fifteen lambs in four years." [3]

Some breeds of sheep, as the Mendip and Dorsets, mentioned by Youatt, breed twice a year, and he gives the following instance in another breed: "In the spring of 1801, Mr. Sheriff, of Kinmyles, Inverness, bought a parcel of ewes in lamb, of the white-faced Highland breed. They lambed in March and April. One old ewe, without a tooth, dropped a second lamb on the 1st of November, 1801, a third on the 29th of April, 1802, and a fourth on the 12th of January, 1803; so that she reared four lambs at different times in the course of twenty-one months." [4]

Rev. Gilbert White gives an account of a half-bred "Bantam" sow that was remarkable for her

[1] "Annual Register," 1806; quoted in Youatt on "Sheep," p. 509.

[2] *Gentleman's Magazine*, March, 1750; quoted in Youatt on "Sheep," p. 509.

[3] *Agricultural Magazine*, April, 1804; quoted in Youatt on "Sheep," p. 509.

[4] *Agricultural Magazine*, February, 1803; quoted in Youatt on "Sheep," p. 509.

fecundity and longevity: "For about ten years this prolific mother produced two litters in the year, of about ten at a time, and once above twenty at a litter; but as there were near double the number of pigs to that of teats, many died. . . . At the age of about fifteen, her litters began to be reduced to four or five, and such a litter she exhibited when in her fatting-pen. . . . At a moderate computation, she was allowed to have been the fruitful parent of three hundred pigs—a prodigious instance of fecundity in so large a quadruped. She was killed in the spring of 1775, when seventeen years old." [1]

A remarkable instance of multiple births is reported in the *Prairie Farmer*, on the authority of the London *Live-Stock Journal*, as follows: "In the neighborhood of Hohenmath, Bavaria, a cow has recently been delivered of five calves at a birth. All of them were born dead, and the mother succumbed a few days later. The calves weighed sixteen, seventeen, eighteen, nineteen, and twenty pounds, respectively, and were all of the same color." [2]

Dr. Simpson states that he has obtained authentic information in regard to forty-two married women who were "born as twins with males," and thirty-six of the number had children. "Two of the females who have families were each born as a triplet with two males." [3]

<hr>

[1] "Natural History of Selborne," p. 222. "The Hog," by Youatt, p. 154.

[2] *Prairie Farmer*, December 8, 1877, p. 389. These cases are, perhaps, not all attributable to heredity. For additional cases of inherited fecundity, *see* p. 16.

[3] "Cyclopædia of Anatomy and Physiology," vol. i., p. 736.

Among cattle, where twin calves are produced, the one a male and the other a female, the latter, called a free-martin, is, as a rule, barren. When the twins are of the same sex, the reproductive powers are not impaired.

In all other varieties of animals, so far as known, when males and females are born together as twins, the females are as prolific as if born singly. In free-martins the internal generative organs are generally imperfect, partaking of the characters of both male and female organs. In appearance these imperfect females frequently resemble steers, the feminine characteristics being mostly wanting.[1]

In rare instances the free-martin is capable of breeding, the reproductive organs not having become malformed from her intra-uterine development with a male.

Youatt, in his work on "Cattle," gives but two cases of fertile free-martins. Dr. Hunter dissected a free-martin calf, that died when a month old, and found the sexual organs naturally constituted, and he also heard of two instances in Scotland of free-martins that were prolific.

Dr. Maulson has likewise published similar cases in London's *Magazine of Natural History.*[2]

A few additional cases might be gathered from the agricultural papers, but they only serve to show that fertility under such conditions is decidedly exceptional.

---

[1] "Cyclopædia of Anatomy and Physiology," vol. ii., pp. 701, 702, 735, 736. Youatt on "Cattle," p. 538.

[2] "Cyclopædia of Anatomy and Physiology," vol. ii., p. 735.

Dr. Simpson, in his valuable paper, from which we have already quoted, says: "As to the cause of the malformation and consequent infecundity of the organs of generation in the free-martin cow, we will not venture to offer any conjecture in explanation of it.

"It appears to be one of the strangest facts in the whole range of teratological science, that the twin existence *in utero* of a male along with a female should entail upon the latter so great a degree of malformation in its sexual organs, and *in its sexual organs only.* The circumstance becomes only the more inexplicable when we consider this physiological law to be confined principally, or entirely, to the cow, and certainly not to hold with regard to sheep, or perhaps any other animal. The curiosity of the fact also becomes heightened and increased when we recollect that when the cow or any other uniparous animal has twins, both of the same sex, as two males or two females, these animals are always both perfectly formed in their sexual organization, and both capable of propagating. In the course of making the preceding inquiries after females born co-twins with males in the human subject, we have had a very great number of cases of purely female and purely male twins mentioned to us, who had grown up and become married; and in only two or three instances, at most, have we heard of an unproductive marriage among such persons. Further, we may in conclusion remark that, among the long list of individual cases of hermaphroditism in the human subject that we have had occasion to cite, we find only one instance in which the malformed being is stated to have been a twin. Katsky, however, Nargele, and Saviard, have each, as before stated, mentioned a case in which both twins were hermaphroditically formed in their sexual organs." [1]

It is worthy of mention, in this connection, that some authors assert that the production of twins in the human species is an abnormal peculiarity, and they claim that a larger proportion of idiots and imbeciles are twin-born than of those not thus affected, that the relatives of imbeciles and idiots frequently have twins, and that, in families where twins are frequently produced, bodily deformities are repeatedly observed.[2]

Dr. Duncan, in his work on " Fecundity, Fertility, and Sterility," is inclined to the belief that the production of twins in the human family is not only abnormal, but that it cannot be relied upon as an indication of great fecundity, as twins are usually produced at longer intervals than single births, and that in the latter the entire period of child-bearing is likely to be more extended.

[1] *Loc. cit.*, p. 786.
[2] *See* Dr. Arthur Mitchell's paper in the *Medical Times and Gazette*, November 15, 1862, referred to in Walford's "Insurance Cyclopædia," vol. iii., p. 192.

CHAPTER IX.

THE term in-and-in breeding is generally used to indicate the breeding together of animals that are closely related.

As to the degree of relationship, in the breeding of animals, to which this term should be applied, it not only appears that no definite rule has been established, but that almost every writer uses it with a different shade of meaning.

The prevailing differences of opinion in regard to the effects of in-and-in breeding have, to some extent at least, arisen from this diversity of meaning in the use of the term, and a misapprehension as to the real advantages that are aimed at in its practice.[1]

---

[1] In-and-in breeding has been defined as follows: " The breeding from close affinities "—Youatt on " Cattle," p. 525.· " The breeding from close relations "—Johnson's " Farmers' Cyclopædia," p. 248. " Breeding between relatives without reference to the degree of consanguinity "—Randall's " Practical Shepherd," p. 116. " It should only be applied to animals of precisely the same blood as own brother and sister "—Bowly, *Journal of the Royal Agricultural Society*, vol. xix., p. 149. " Breeding from the same family, or putting animals of the nearest relationship together "—Sinclair's " Code of Agriculture," p. 93. " The pairing of relations within the degree of second cousins, twice or more in succession "—Stonehenge on " The Horse," p. 140.

7

Sir John Sebright, whose successful practice gives the weight of authority to his opinions on this subject, is often quoted as an opponent of in-and-in breeding. He evidently, however, limits the application of the term to the frequent repetition of the closest relationship in parents. He says: "Mr. Meynel's foxhounds are likewise quoted as an instance of the success of this practice; but, upon speaking to that gentleman upon the subject, I found that he did not attach the meaning that I do to the term in-and-in. He said that he frequently bred from the father and the daughter, and the mother and son. That is not what I consider as breeding in-and-in, for the daughter is only half of the same blood as the father, and will probably partake, in a great degree, of the properties of the mother.

"Mr. Meynel sometimes bred from brother and sister; this, certainly, is what may be called a *little close;* but should they *both be very good*, and particularly should the same defects not predominate in both, but the perfections of the one promise to correct in the produce the imperfections of the other, *I do not think it objectionable:* much further than this, the system of breeding from the same family cannot, in my opinion, be pursued with safety." [1]

He then proceeds to point out the difficulties that arise, in the practice of what he calls in-and-in breeding, from the rare instances in which breeding-animals are found to be free from defects.

If the terms inbreeding, close breeding, and interbreeding, are used to indicate the breeding to-

---

[1] "The Art of improving the Breed of Domestic Animals," pp. 8, 9.

gether of closely-related animals in a single instance, or at long-separated intervals, the term in-and-in breeding could then be used with greater exactness to indicate the frequent repetition of the process.

High breeding implies a careful selection of breeding-animals within the limits of a family, with reference to a particular type, and regardless of relationships. High-bred animals are not necessarily in-and-in bred, although, from the system of selection practised, they must be closely bred to a greater or less extent.

The opponents of in-and-in breeding claim that it produces a delicacy of constitution—a predisposition to disease, and a lack of fecundity—and they often fall into the error of assuming that all who do not admit the truth of these claims are in favor of close breeding as a rule of practice.

For the purpose of gaining a knowledge of the principles involved in the breeding together of animals that are closely related, we will first examine the practice of those who have gained a high reputation as breeders of domestic animals, and then consider the objections to their methods of improvement.

From the time of Bakewell, the breeders who have gained the greatest reputation have evidently aimed to establish in their flocks and herds certain well-marked characters that adapted the animal to a particular purpose.

In giving expression to their ideal type, or standard of excellence, they found it necessary to limit their selection of breeding-stock to the animals that had the characters they wished to perpetuate. As it was only among the animals descended from a common ancestry—with the same hereditary tendencies—that the desired variations were found, they were frequently compelled to breed together animals that were more or less closely related.

Their selections were made to secure in both parents the same general characteristics that they wished to obtain in their offspring, and the close relationships observed in their breeding-stock were but the necessary *incidents* of their practice.

Close breeding with them was but a *means* of improvement, and not an *end* that was thought to be desirable in itself.

The true method of improvement practised by these eminent breeders is frequently misunderstood, and their intentions have, consequently, been misrepresented. A friend of mine, on his return from England, told me that he had learned an important *secret* in breeding that he believed to be a prevailing rule among the best breeders. It was this: "Breed from half brother and sister;" and an examination of a large number of the most celebrated pedigrees apparently made the theory a plausible one.

An incidental feature in the methodical improvement of animals had, however, been mistaken for the real causes of improvement, which were entirely overlooked.

Animals are not *improved* by breeding except in the increased stability gained in dominant characters, and the certainty with which they are transmitted, as the offspring, at the time of birth, can only be possessed of the characters they have derived from their ancestors. The true means of improvement have already been pointed out in the chapter on variation, and we must look upon methods of breeding solely with reference to the perpetuation of characters thus obtained.

No matter what opinions we may form as to the advantages or disadvantages of close breeding, the fact remains the same: that all the great breeders have practised it to a greater or less extent, and, as far as we are able to judge, with the same purpose—that of retaining and fixing in their flocks and herds certain desirable characters that have been developed by modified conditions.

The extent to which in-and-in breeding has been practised by the breeders who have attained the greatest celebrity is shown in the accompanying diagrams.[1]

The pedigrees in a number of the diagrams are arranged on a new plan; the name of each animal being given but once, while the lines are drawn so that the relationships can be readily traced.

The pedigrees are selected to represent the most popular families of the leading breeds, as indicating

[1] Some of these diagrams were prepared to illustrate a lecture on "In-and-in Breeding," delivered by the author before the "American Association of Breeders of Short-Horns," at their meeting in Cincinnati, December 3, 1873, and published in their Transactions. The lecture and diagram were also published in the "Report of the Michigan State Board of Agriculture for 1872." The illustrations have been made use of in this chapter, but the matter has been entirely rewritten and arranged to conform to the classification of topics in this work.

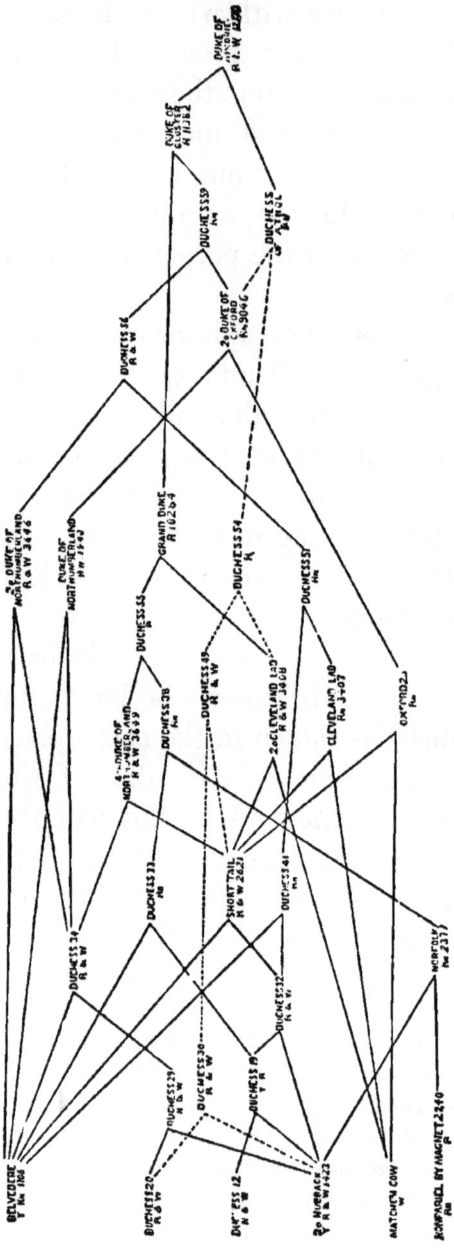

DIAGRAM 1.—PEDIGREE OF DUKE OF AIRDRIE (12730).

the uniformity of the practice in the improvement of all classes of animals.

In all of the cases cited, the breeders evidently intended *to breed together animals of the same qualities, regardless of relationship.*

In Diagram 1, the pedigree of Duke of Airdrie may be traced, beginning with the six animals at the extreme left of the diagram.

In Diagram 2, the pedigrees of these same six animals, arranged in a different order, may be traced back to Favorite by black lines, while the dotted lines if continued would run to Hubback.

Certain animals in Diagram 1 are arranged in a different order, and taken for the basis of Diagram 3, which gives the pedigrees of some of the "New York Mills" herd.

Diagram 8 gives the pedigree of a number of animals bred by Charles and Robert Colling.

In all of the diagrams, the two lines coming together at the left of a name trace respectively to the sire and dam, while the lines from the right of the name run to the offspring.

The Booths practised close breeding to a great extent, as will be seen from an examination of the pedigree of their most noted animals. The pedigree of the sisters Queen of the May, Queen Mab, Queen of the Vale, and Queen of the Ocean; and their brother, Lord of the Valley (14837), bred by R. Booth, of Warlaby, is given in Diagram 4, on page 147.

According to the calculations of Rev. J. Stone, of Hellidon, "Crown Prince is 1055 times descended

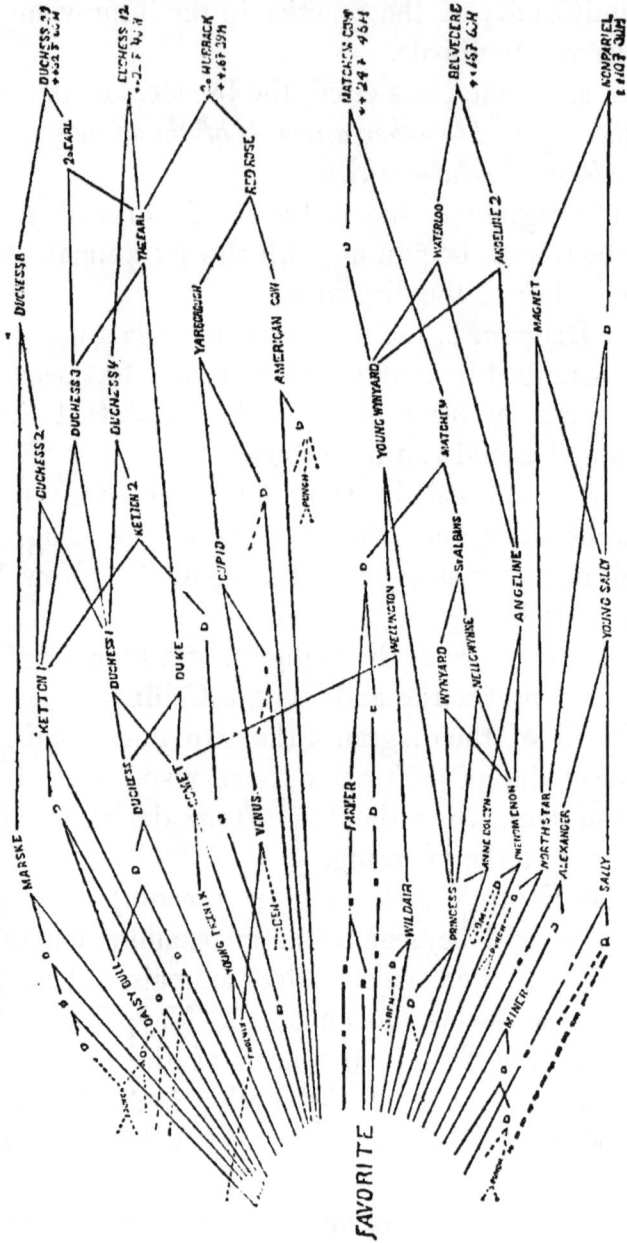

DIAGRAM 2.—PEDIGREE OF "BATES" SHORT-HORNS.

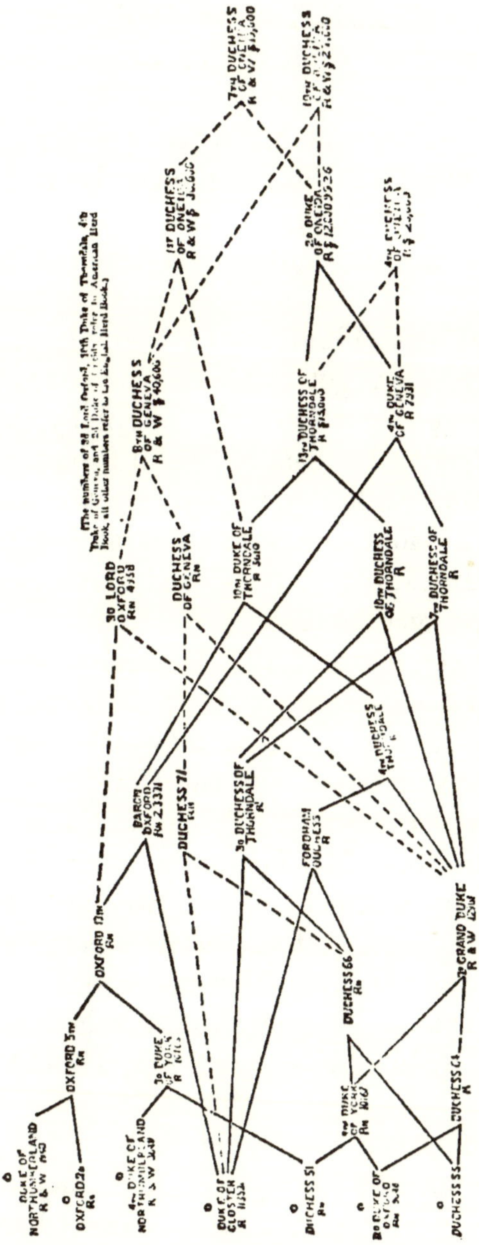

DIAGRAM 3.—PEDIGREE OF "NEW YORK MILLS" SHORT-HORNS.

from Favorite, and Red Rose by Harbinger, 1344 times. So the produce of the two are descended from him 2399 times." [1]

Lord of the Valley and his sisters are three sixteenths of the blood of Pilot, although he had not been used for five generations.

They are also five-sixteenths of the blood of Buckingham.

Lord of the Isles (18267) is an example of still closer breeding; he was got by Sir Samuel out of Red Rose, by Harbinger. Sir Samuel was got by Crown Prince out of Charity, the dam of Crown Prince. According to the same authority, the *out-crosses* made use of by Mr. Booth had a strong infusion of the blood of Favorite, Mussulman having sixty-four crosses, Lord Lieutenant one hundred and six crosses, and Matchem fifty-two crosses of this favorite progenitor of the improved Short-Horns.

Diagram 7 shows the extent to which in-and-in breeding has been practised with the Herefords.

Mr. Price, the celebrated breeder of Herefords, says: "I bought from Mr. Tompkins a considerable number of his cows and heifers, and two more bulls. I have kept the blood of these cattle unadulterated for forty years, and Mr. Tompkins assured me that he had bred the whole of his stock from two heifers and a bull, selected by himself early in life, without any cross of blood.

"My herd of cattle has, therefore, been bred in-and-in, as it is termed, for upward of eighty years, and by far the greater part of it in a direct line, on

[1] Carr's "History," p. 40.

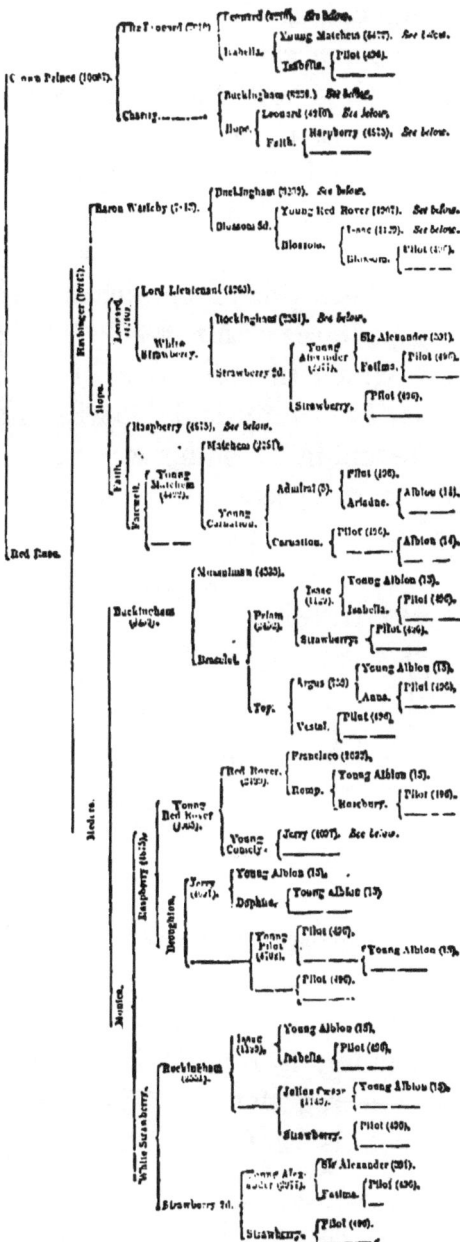

DIAGRAM 4.—PEDIGREE OF "BOOTH" SHORT-HORNS.

both sides, from one cow now in calf for the twentieth time. I have bred three calves from her, by two of her sons, one of which is now the largest cow I have, possessing also the best form and constitution; the other two were bulls, and proved of great value, thus showing indisputably that it is *not* requisite to mix the blood of the different kinds of the same race of animals, in order to keep them from degenerating." [1]

The following pedigree of Mr. Fowler's celebrated bull Shakespeare, which includes that of Mr. Bakewell's noted bulls Twopenny and D, will show the extent to which in-and-in breeding was practised by those who were most successful in improving the Long-Horn breed :

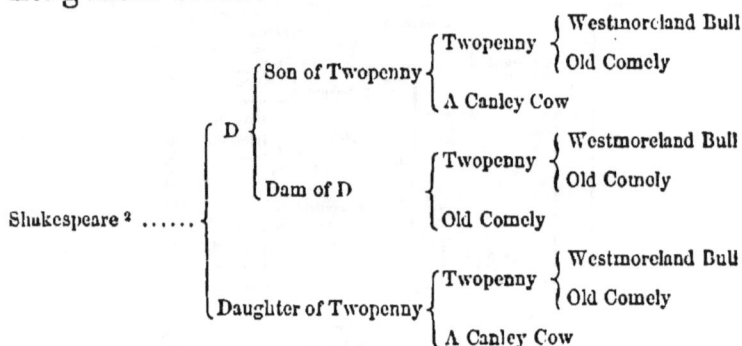

```
                                       ⎧ Westmoreland Bull
                        ⎧ Twopenny  ⎨
           ⎧ Son of Twopenny ⎨        ⎩ Old Comely
           ⎪            ⎩ A Canley Cow
      ⎧ D ⎨
      ⎪    ⎪                         ⎧ Westmoreland Bull
      ⎪    ⎪            ⎧ Twopenny ⎨
      ⎪    ⎩ Dam of D ⎨           ⎩ Old Comoly
Shakespeare [2] ......⎨            ⎩ Old Comely
      ⎪
      ⎪                            ⎧ Westmoreland Bull
      ⎪               ⎧ Twopenny ⎨
      ⎩ Daughter of Twopenny ⎨    ⎩ Old Comely
                              ⎩ A Canley Cow
```

Mr. Quartly, the great improver of the Devons, bred his animals very closely.

The name of his bull, Prince of Wales (105), a celebrated prize-winner, is repeatedly found in the pedigrees of the best-bred Devons of the present day. His pedigree, which is as follows, shows that half

[1] *Farmer's Magazine*, 1841, vol. xiv., p. 50.

[2] Marshall's "Midland Counties," vol. i., pp. 320-322 ; Youatt on 'Cattle,' pp. 192, 193 ; Low's "Domestic Animals," p. 375.

brother and sister were bred together twice in succession:

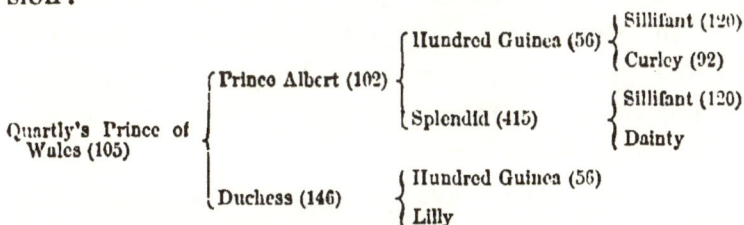

```
                                              ⎧ Sillifant (120)
                          ⎧ Hundred Guinea (56) ⎨
                          ⎪                    ⎩ Curley (92)
              ⎧ Prince Albert (102) ⎨
              ⎪           ⎪                    ⎧ Sillifant (120)
Quartly's Prince of ⎨    ⎩ Splendid (415)      ⎨
   Wales (105)     ⎪                           ⎩ Dainty
              ⎪                    ⎧ Hundred Guinea (56)
              ⎩ Duchess (146)      ⎨
                                   ⎩ Lilly
```

The high-bred cow, Eveleen 5th (466), belonging to the Michigan State Agricultural College, is a regular breeder, a good milker, and remarkable for her feeding qualities and sound constitution.

Her dam traces to Forester thirty times, to Sillifant ten times, to Hundred Guinea seven times, and to Quartly's Prince of Wales three times, in eight generations.

Her sire, within the same limits, traces to Forester twenty-one times, to Sillifant twenty times, to Hundred Guinea thirteen times, and to Quartly's Prince of Wales five times.

In-and-in breeding has not been practised to the same extent with horses as with other farm-stock, yet many of the most noted horses on record have been bred from close relationships.

Stonehenge says: "When any new breed of animals is first introduced into this country, in-and-in breeding can scarcely be avoided; and hence, when first the value of the Arab was generally recognized, the breeder of the race-horse of those days could not well avoid having recourse to the plan. Thus we find, in the early pages of the stud-book, constant instances of very close breeding, often carried to such an extent as to become incestuous." And he adds,

"The evidence of success in resorting to the practice of in-breeding is too strong to be gainsaid."[1]

The pedigree of Goldsmith's Maid, the "queen of the American turf," is given in Diagram 5, as an illustration of a well-bred trotter, as it includes several other noted pedigrees.

Sheep-breeders have quite generally practised in-and-in breeding with the best results.

The merino sheep, bred by the late Edwin Hammond, present a remarkable example of close breeding. "They were bred in-and-in by Colonel Humphreys up to the period of Mr. Atwood's purchase; Mr. Atwood bred his entire flock from *one ewe*, and never used any but pure Humphreys rams; Mr. Hammond has preserved the same blood entirely intact, and thus, after being drawn beyond all doubt from an unmixed Spanish Cabana, they have been bred in-and-in, in the United States, for upward of sixty years."[2]

The pedigree of the ram Gold Drop, for which Mr. Hammond refused twenty-five thousand dollars, is given in Diagram 6, in convenient form for study. Dividing the "*blood*" of Gold Drop into 512 parts, it would be made up as follows:

	Parts.
Of Old Black	196
" first choice of old ewes . . .	151
" " " " " ewe-lambs . .	109
" dam of light-colored ewe . . .	28
" Old Matchless	28
Total	512

[1] Stonehenge on the "Horse," pp. 140, 141. *See* also "British Rural Sports," pp. 422–425, 286.

[2] Randall's "Practical Shepherd," p. 120.

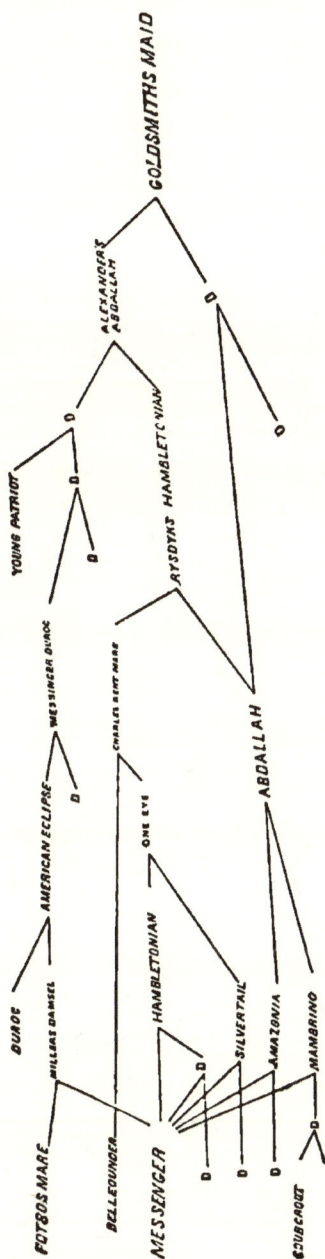

DIAGRAM 5.—PEDIGREE OF GOLDSMITH'S MAID.

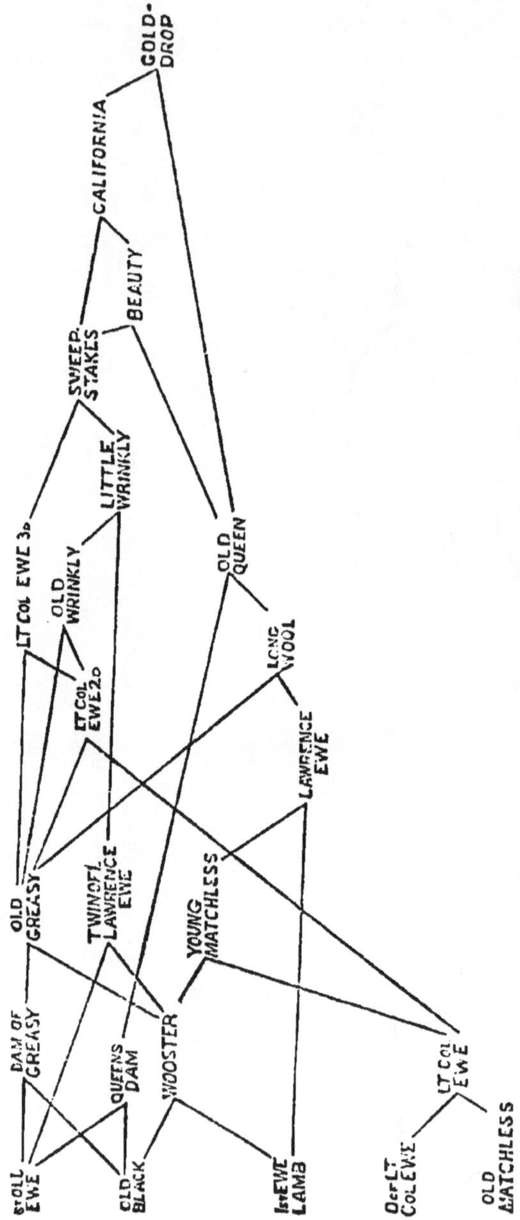

DIAGRAM 6.—PEDIGREE OF EDWIN HAMMOND'S MERINO RAM GOLD DROP.

Old Greasy also represents 188 parts of blood in 512, and Wooster represents 138 parts in 512. Sweepstakes is $\frac{9}{16}$ of the "blood" of Old Greasy, and $\frac{13}{32}$ of the "blood" of Wooster.

It will be seen that ten lines of descent may be traced from Old Greasy to Gold Drop, and fourteen lines of descent may be traced from Wooster to Gold Drop.

The Rich family of merino sheep furnish another example of successful close breeding.

Mr. Randall says they "were first crossed in 1842. They were then preëminently hardy. No one claims that they have gained either in hardiness or size by the cross; yet for thirty years preceding that period they had been bred strictly in-and-in, to say nothing of their previous in-and-in breeding in Spain."[1]

"The Messrs. Brown during fifty years have never infused fresh blood into their excellent flock of Leicesters.

"Since 1810 Mr. Barford has acted on the same principle with the Foscote flock.

"He asserts that half a century of experience has convinced him that when two nearly-related animals are quite sound in constitution, in-and-in breeding does not induce degeneracy, but he adds that he 'does not pride himself on breeding from the nearest affinities.'"[2]

From the examples that have thus far been pre-

[1] Randall's "Practical Shepherd," p. 119.

[2] Darwin's "Animals and Plants under Domestication," vol. ii., p. 149.

sented, it appears that in-and-in breeding has been
quite generally practised by those who have been the
most successful in improving the different breeds, and
it is probable, to say the least, that they have all made
use of it with a common purpose.

If those having the greatest reputation in the art
had resorted to the practice of in-and-in breeding, on
account of the direct influence it was in itself supposed to exert in the improvement of animals, they
would undoubtedly have made use of it to a greater
extent than they have done.

From a careful examination of the pedigrees we
have quoted, or any others that may be found in the
herd-books and breeding-registers, representing the
practice of breeders of acknowledged reputation, it
will be found that in-and-in breeding has only been
resorted to in the case of some favorite animal or animals that were superior in certain respects to the
average members of the herd or family which they
represent, and the object has evidently been to secure, in their offspring, a predominance of their most
highly-valued characters.

From the complex relations of the multitude of
hereditary characters in animals, which have been
derived, as we have seen, from all of their ancestors,
the modifying influences of food and habit cannot
affect all animals in precisely the same degree or
manner, and they cannot, therefore, be expected to
produce the same modifications in the characters of
a large number of animals at the same time. The
breeder who makes an intelligent use of these modifying agencies, in the improvement of his stock, will

rarely find more than one or two animals presenting variations that approximate closely to the ideal standard of excellence he has adopted; and, moreover, it must be admitted that the same desired form of variation will be more likely to be obtained in those animals that have the closest resemblance in their hereditary tendencies and constitution. The truth of this proposition is amply proved by experience, as we find that the desirable variations that have laid the foundation for the improvement of breeds, have, as a rule, occurred in a few favored animals, belonging to the same family, and closely related in blood.

Distinct breeds of animals have originated, as we have seen, through the influence of the conditions to which they were subjected in particular localities, in connection with a continued selection of those that in their form and qualities resemble each other, while those presenting diverging characters were rejected.

In the *improved* breeds we have ingrafted upon the original type the highly-artificial characters that render the animal valuable for a special purpose, and these, from their very nature, are more difficult to retain than the less divergent characters of the original breed.

These artificial characters can only be secured, in their greatest perfection, by persistent effort in the systematic accumulation of slight variations in the desired direction, and they can only be made the dominant characters of a family or breed by breeding exclusively from those animals in which they are the most conspicuous.

If, as may reasonably be expected, these characters

make their appearance in a single family, or in a few individuals that are closely related, in-and-in breeding must necessarily be resorted to to secure their perpetuity. From these considerations it must be obvious that, in the improvement of a breed, in-and-in breeding tends to produce uniformity in the characteristics of a family by fixing desirable variations and making them dominant.

From the uniformity thus obtained in the hereditary tendencies of the organization—the dominant characters of all the immediate ancestors being the same—the power of hereditary transmission is likewise increased, as observed in what is now called prepotency. If the hereditary transmission of desirable variations were not intensified by the process of inbreeding, or otherwise, they would unavoidably become latent by the preponderance of the more stable characters of the original type.[1]

[1] The difficulty of fixing a particular variation that presents a marked divergence from the normal condition of an animal, may be illustrated as follows: If we suppose the hereditary tendencies of the animal to be represented by one hundred and the desired variation by one, the chances of its being perpetuated by the animal when bred with another, that had not an equal susceptibility to variation in the same direction, would be only one in one hundred under the most favorable conditions; but when there is a tendency to the dominance of other characters, the chances of its repetition will be less. Or, if among one thousand animals of a given breed there are but two that present a slight variation of a particular character, the chances of its being preserved, if the animals in which it occurs are not bred together, would evidently be but two in one thousand, even in case the variation was not more difficult to preserve than the ordinary characters, while in the case of a variation of a highly-artificial character that would be likely to be obscured by more stable characters, the chances of its preservation would be materially diminished.

Such characters, if not inbred, might be inherited, and make their appearance at intervals, as in the case of what are called accidental characters, which are not likely to be transmitted in a dominant form, but they would not become family characteristics.

In the breeding of animals, the parent that apparently exercises the greatest influence upon the dominant characters of the offspring is said to be prepotent.

When certain desirable characters have been developed, in a few individuals, they can only be ingrafted upon the entire flock, or herd, by making them the dominant characters of the males that are to be used, and securing in them prepotency in their transmission.

As the male practically represents one-half of the breeding flock or herd, prepotency in the transmission of his better qualities is one of the most valuable characteristics he can possess.

In the cases of marked prepotency, in which the ancestral history of the animals can be traced, in-and-in breeding has been so generally practised that we cannot avoid the conclusion that the one is dependent upon the other.

If the male is more highly-bred than the females with which he is coupled, a greater uniformity in the offspring will be obtained through the predominance of his characteristics.

The great demand for high-bred males, by the best breeders of all classes of stock, is the cause of the prevailing high prices of animals belonging to the most fashionable families.

The importance of securing prepotency in the male parent has apparently been recognized by all the great breeders, as we find, as a rule, that their breeding-males have been selected from sub-families that are more highly inbred than the average of their stock. In many cases the practice of in-and-in breeding has been limited to certain families that were set apart for breeding-sires, and this on many accounts would undoubtedly be the best method.

Mr. Hammond's "Queen" family, from which he selected his rams, were bred in-and-in to a greater extent than the rest of his flock. Jonas Webb kept five separate flocks, the rams used by himself being drawn from his favorite family; the "Duchess" tribe was the source of the sires of Mr. Bates's herd, and Mr. Booth had his favorite families, from which the sires of his own herd are descended.

The degree of high breeding required to secure prepotency in a given male will evidently depend upon the relative development and breeding of the females with which he is coupled; the better the females, and the greater the uniformity in their characteristics, the more intense must be the power of transmission in the male to secure a predominance of his peculiarities in his offspring, and this intensity in the power of transmission can only be produced by still higher breeding.

The supposed cases of spontaneous prepotency and accidental variation cannot reasonably be claimed to constitute exceptions to the generally acknowledged laws that determine variations and regulate their transmission, as they are readily explained when all of the facts relating to them can be ascertained.

As an illustration of this class of cases, attention is called to the family history of Mr. Fowler's Long-Horn bull Shakespeare, which is often quoted as an example of spontaneity. His pedigree, as given on page 148, shows him to have been deeply in-and-in bred from animals of acknowledged merit. Mr. Marshall says: "This bull is a striking specimen of what naturalists term *accidental varieties*.

"Though bred in the manner that has been mentioned, he scarcely inherits a single point of the Long-Horned breed, his horns excepted." [1]

Then follows a description of the animal that shows him to have been somewhat better, in general form, than the ordinary Long-Horns of his day. [2]

Mr. Marshall also says, Mr. Fowler's "cows have long been considered of the first quality—of the best Canley blood—and his bull Shakespeare, already mentioned, has raised them to a degree of perfection, which, in the opinion of the first judges, the breed of cattle under notice never before attained." [3]

[1] "Rural Economy of the Midland Counties," vol. i., p. 322.

[2] *Loc. cit.*, p. 322.

[3] A careful comparison of Mr. Marshall's description of the bull Shakespeare with his "general description" of "the higher class of individuals" of the Long-Horn breed, in the herds of Messrs. Fowler, Bakewell, and Princep, will show that aside from a deeper chest, shorter legs, and a peculiarity in the setting on of the tail, Shakespeare did not differ essentially from the best type of the breed to which he belonged.

In his "general description" of the breed, Mr. Marshall says, "The tail set on variously, even in individuals of the highest repute." So that the peculiarities in the tail of Shakespeare cannot be considered as exceptional in a breed in which variety in the character was the rule.

Compare Marshall, *loc. cit.*, pp. 323–325, with pp. 327–331 ; and Youatt on "Cattle," pp. 193–197.

The Canley stock, which was the foundation of Mr. Fowler's herd, as well as Mr. Bakewell's, was obtained of Mr. Webster, of Canley, "the leading breeder of the midland counties."

Of this stock Mr. Marshall says: "I have, indeed, heard it said, by a man who has himself been a breeder of some eminence, that Mr. Webster had the best stock, especially of *beance* (cattle), that ever were, or (he believed) ever will be bred in the kingdom;" and he adds in a note, "Another eminent breeder, on whose judgment I can better rely, is of opinion that, in *beauty* or *utility* of form, they have received little, if any, improvement since Mr. Webster's day." [1]

Old Comely (the dam of Twopenny) was killed at the age of twenty-six years, and "the fat on her sirloin was four inches in thickness." [2]

The bull Twopenny was a celebrated animal, and the bull D, Mr. Marshall says, was "a fine animal, and a striking proof of the vulgar error that breeding in-and-in weakens the breed. . . . At the age of twelve or thirteen years (he) is more active and higher-mettled than bulls in general are at three or four years old." [3]

From what is known of the ancestors of the bull Shakespeare, his superior qualities could not have been accidental; and, as the progeny that he left were unmistakably Long-Horns of the most approved type, he must have transmitted the characters he inherited from his ancestors.

[1] *Loc. cit.*, p. 319.
[2] *Farmer's Magazine*, vol. xvii., p. 84.
[3] *Loc. cit.*, p. 321.

If he presented any characters that were not to be reconciled with the improved type of his breed—as in the fancied resemblance to "a Holderness or Teeswater bull"—it would certainly be more reasonable to refer them to the atavic transmission of some remote cross than to accidental variation.

In attempts to ingraft a new or modified character upon those representing a family type, without destroying the specific characters of the family, close breeding within the limits of the family must be practised to prevent too wide a divergence in the dominant characters.

Diagram 6 may be studied with profit, as it shows the skillful manner in which the blood of the light-colored ewe was infused into the flock of Mr. Hammond, to tone down the tendency to the production of excessive yolk without destroying the other desirable qualities of the descendants of old Black, and the first choice of old ewes and first choice of ewe-lambs.

From this general examination of the practice of in-and-in breeding by the most celebrated breeders, it appears that they have made use of it to secure uniformity in their breeding-stock, to fix the slight variations that they sought in the process of improvement and blend them with the best original characters, and to secure the important quality of prepotency in the males that they made use of to "improve" the average characters of their stock.

We will now proceed to a consideration of the alleged influence of in-and-in breeding in producing delicacy of constitution, lack of fecundity, and a tendency to disease and abnormal peculiarities. It is, without question, too often the case, that high-bred animals have one or more of the defects in question, to an extent that seriously impair their value for any useful purpose; and it is undoubtedly the interest of the breeder to ascertain the true causes of their prevalence, and the best methods of counteracting them.

If an imaginary cause is mistaken for the real one, the breeder may, by avoiding it, rest in fancied security, while the unsuspected agencies that he has overlooked may be acting with undiminished energy. We have already observed that, in the highest development of special characters by artificial treatment, particularly in the meat-producing breeds, a delicacy of constitution is produced that renders the animal more susceptible to the influence of modifying agencies.

When this impressibility of the organization is in excess and becomes a marked characteristic of a family, it will be fixed, and perhaps intensified, by in-and-in breeding; or, in other words, if a delicacy of constitution is produced by the system of management to which animals are subjected, it will readily be made a prominent characteristic by in-and-in breeding.

That the close breeding in this case is not the cause of the impaired condition of the organization, but rather the means of its being perpetuated, cannot be doubted. The following cases, in connection with a number of a similar character that have already been cited, will show that in-and-in breeding is not necessarily associated with a delicacy of constitution, and it does not, as a matter of course, produce it.

In the first volume of the "Hereford Herd-Book"

is a portrait of the closely in-and-in bred bull Cotmore (376): "He was the winner of the first prize in his class at the first meeting of the Royal Agricultural Society of England, held at Oxford. He was also a winner of many local prizes, and was, perhaps, one of the finest bulls ever seen; his colossal proportions were something very astounding, as may be inferred from the fact that the live weight was thirty-five cwt.

"He was bred by Mr. Jeffries, of the Grove, near Leominster."

The pedigree of Cotmore, given in Diagram 7, shows the closest in-and-in breeding.

"Sovereign (404), when at the age of *fifteen* years, was his sire, but he was not of the same enormous size, although acknowledged to be one of the best stock-getters of his day. He was bred by Mr. Fewer, and very closely in-and-in bred." [1]

The pedigree shows that the sire of Cotmore (Sovereign) was the produce of (Favorite and Countess) full brother and sister, their sire and dam (Young Wellington and Cherry) were half brother and sister, and their grandams (Silky and Old Cherry) were half-sisters. Lottery, the sire of Cotmore's dam, was not only closely bred, but we find him descended from the same animals as Sovereign. The pedigree of Cotmore's grandam is not given in the "Herd-Book."

Mr. George Butts, of Manlius, New York, has recently furnished an instance of continued close breeding in his family of Short-Horns. He says: "I bred Apricot's Gloster 2500 upon the second generation

---

[1] "Hereford Breed of Cattle," by T. Duckham, p. 18, in vol. vi. of "Hereford Herd-Book."

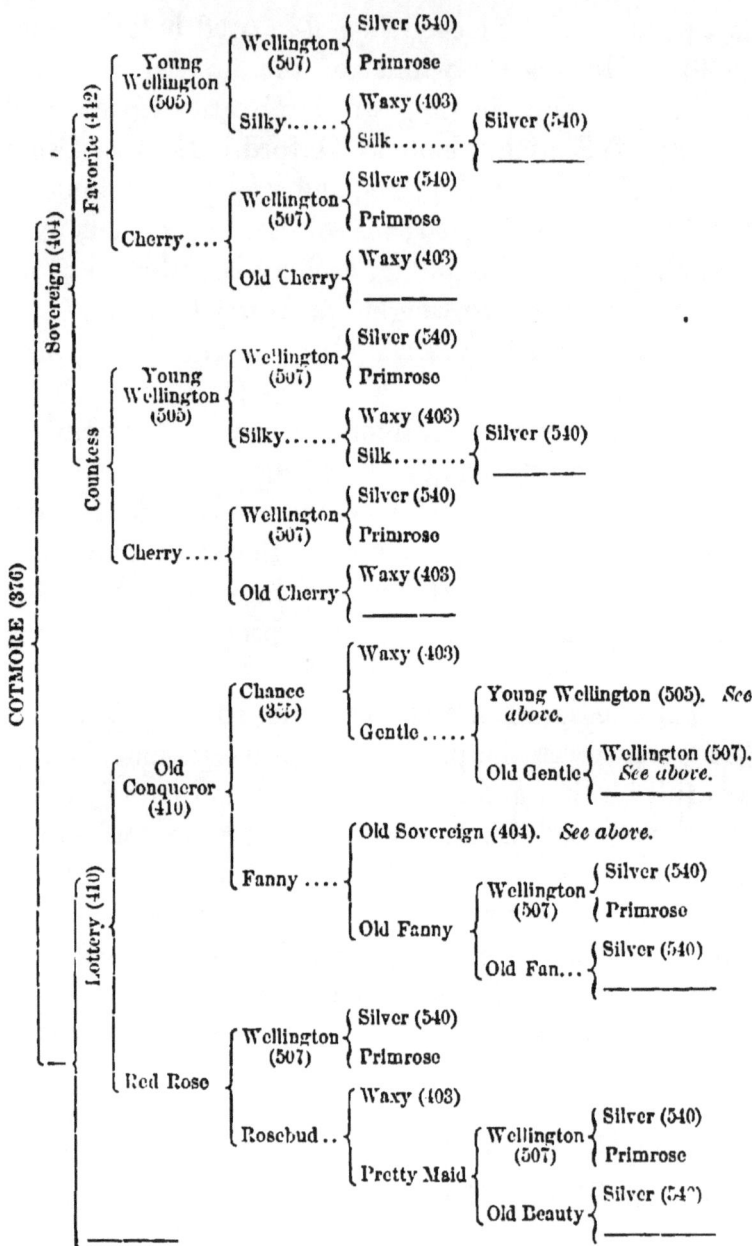

DIAGRAM 7.—PEDIGREE OF HEREFORD BULL COTMORE.

of heifers of his own get, thereby producing Treble
Gloster 7331. I then bred him (Treble Gloster)
back to his dam, Spring Beauty, and the result was a
very fine heifer, Souvenir. I then bred Treble Gloster again to Souvenir, and the result was an extra fine
heifer which is May Beauty; and I wish here to say
that the results of the above course of breeding have
been so entirely satisfactory in the past, that I am
now breeding Treble Gloster to all my females, regardless of his relationship to them, in the fullest
confidence in this course of breeding." [1]

It should be remarked that such practice is not to
be recommended, except in cases where there is some
special object to be accomplished, on account of the
difficulty of finding animals free from defects.

The success of Mr. Butts, thus far, shows that the
stamina of animals is not necessarily impaired by the
closest possible breeding.

The wild cattle of Chillingham Park, in England,
have, as is well known, been bred within the limits of
the herd for many years, their origin and the time of
their inclosure in this park being unknown. Lord
Tankerville, in 1838, said, "In my father and grandfather's time, we know the same obscurity as to their
origin prevailed." [2]

Mr. Darwin states that "the late Lord Tankerville
owned that they were bad breeders," [3] and he estimated
the increase of the herd in 1861 at about one in five.

---

[1] *Country Gentleman*, 1874, p. 409.

[2] *Farmer's Magazine*, vol. xxxvi., p. 354.

[3] "Report of the British Association," 1838, quoted in "Animals
and Plants under Domestication," vol. ii., p. 148.

When I saw the herd, in 1874, it numbered about sixty, of all ages and sexes. Among them were several steers.

The park-keeper informed me that they produced from ten to twelve calves annually, which agrees closely with Mr. Darwin's estimate. They are certainly not very prolific, yet the number of calves is, perhaps, as great as could be expected under the conditions in which they are placed.

They exhibited no indications of degeneracy or lack of constitutional vigor, and I was assured that they were both healthy and hardy. After several hundred years of close breeding they are apparently as robust as animals that have frequently received infusions of "new blood" by crossing.

Mr. Ballance, who has bred Malay fowls for nearly thirty years, says : "During the whole of this period I have never allowed the introduction of any fresh blood by crossing with any other strain of Malays, but have kept entirely to my own; and as I have succeeded in winning more prizes with Malays than any other fancier of these much-abused but most valuable birds, in all parts of the kingdom, I think my experience is not to be despised, as testifying to the fact that breeding in-and-in does not necessarily deteriorate the birds who may be subjected to this operation." [1]

"Colonel Jaques, of the Ten Hills Farm, near Boston, imported a pair of Bremen geese in 1822. They were bred together till 1830, when the gander was accidentally killed.

[1] Tegetmeier's "Poultry-Book," p. 79.

"Since then the goose bred with her offspring, till she was killed by an attack of dogs in 1852. Great numbers were bred during this time, and of course there was much of the closest breeding, yet there was no deterioration, and in fact some of the later ones were larger and better than the first pair. The same gentleman also obtained a pair of wild-geese from Canada in 1818, which, with their progeny, were bred from, without change, until destroyed by dogs with the above-named in 1852. They continued perfect as at first."[1]

Mr. James Ruthven, formerly secretary of the North British Columbarian Society, says: "There is one fact I became acquainted with three years since. A gentleman in Ireland got one pair of trumpeter pigeons, and put them into a large loft alone. He kept them there *fifteen years*, breeding and producing, without once adding fresh blood; only, when they got too numerous, killing off. The produce are as strong birds and as healthy as could be desired."[2]

Mr. Dixon, of Canandaigua, New York, says he has bred Dominique fowls for twelve years without having a "strange cock in the yard during that time;" and his stock is strong and hardy, the hens laying as well as those of his neighbors who change their stock often. The cocks averaged about seven pounds and the hens about four pounds."[3]

The fecundity of animals, as has been shown in a preceding chapter, is determined by inheritance, and

[1] Goodale's "Principles of Breeding," pp. 99, 100.
[2] Wright's "Book of Poultry," p. 295.
[3] *Country Gentleman*, February, 1868, p. 112.

the action of the various modifying influences to which they are subjected. In the cases of impaired fecundity in animals that have been bred in-and-in, it will be necessary to ascertain the extent to which these obvious causes of a defective performance of the reproductive functions are operative, before we are justified in assuming the existence of some "occult" or mysterious influence arising from consanguinity.

Mr. John Wright, who is often quoted as an opponent of in-and-in breeding, has evidently overlooked the existence of the more obvious causes of sterility and barrenness, and assumed that they are produced only by close breeding.[1]

He apparently concludes that, when two facts are associated in a large number of cases, they must have the relation of cause and effect. The most striking case cited in support of his theory is as follows:

"In pigs, the writer's experience was considerable, in breeding from three or four sows at the same time, all descended from the same parents, boar and sow; these were put to the same boar for seven descents or generations; the result was, that in many instances they *failed to breed*, in others they bred few that lived; many of them were idiots—had not sense to suck; and, when attempting to walk, they could not go straight. The last two sows of the breed were sent to other boars, and *produced several litters* of healthy pigs. In justice to the advocates of the in-and-in principle, it is but right to state that the best sow during the seven generations was one of the last descent. She was the only pig of that litter. She would not breed

[1] *Journal of the Royal Agricultural Society*, vol. vii., pp. 204, 205.

to her sire, but bred to a stranger in blood at the first trial. She possessed great substance and constitution and was a very superior animal."

It would be difficult to find a better illustration of the physiological principles that have been already presented, and, as they furnish a satisfactory explanation of all the observed facts, we cannot with reason attribute them to the influence of other causes. An acquaintance of Mr. Wright's says of his pigs, "They are of excellent quality, readily feed, and soon attain maturity." [1]

The fattening qualities of these pigs had been highly developed, and finally became a dominant characteristic, while the procreative powers were made latent.

In successfully gaining a single character, Mr. Wright had neglected another essential quality that was obscured for the time being. The high development of the fatty diathesis would be sufficient to account for the lack of fecundity observed, even if there was not on the start an hereditary tendency in the same direction.

That the procreative powers were not destroyed, but remained latent, is shown by the fact that the sows bred freely with boars of another family.

With boars of their own blood they could not be expected to breed, as the powers of fecundity, in such case, would be latent in both male and female, but, when they were bred with animals in which the reproductive function was not latent, the defect was corrected.

[1] *Farmer's Magazine*, vol. xxxvi., p. 388.

It has been supposed by some writers that, in cases like the preceding, where high-bred animals do not breed readily among themselves, a condition exists analogous to that observed in "self-impotent" plants and other hermaphrodite organizations that are incapable of self-fertilization.[1] The cases, however, in which high-bred animals are perfectly prolific seem to indicate that in-and-in breeding and fecundity are not incompatible, and that the loss of fertility, when it occurs in high-bred animals, is better accounted for on the principle that, in the correlation of functions, if one is greatly in excess, another may be obscured.

As the impaired function of reproduction may frequently be restored by a suitable selection of animals within the limits of a high-bred family, it would likewise appear that the functional defect of the organization is not a specific one resulting from an approximate identity in blood.

From the well-known fact that high-bred animals, when kept under different conditions are more prolific than those that are treated in the same manner,[2] it must be apparent that the suspension of the reproductive functions, in the cases under consideration, is produced by the modifying agencies to which the animals are subjected, and not by close breeding.

The infertility of some of the Booth family of Short-Horns[3] has been attributed to the forcing sys-

---

[1] Darwin's "Animals and Plants under Domestication," vol. ii., p. 177.

[2] Sebright's "Art of improving the Breeds," etc., p. 16 ; "American Cattle," by Allen, p. 206 ; Sinclair's "Code of Agriculture," p. 95.

[3] Carr's "History," p. 90.

tem practised in "training" for exhibition, rather than in-and-in breeding, and this is undoubtedly a potent cause of sterility and barrenness.

In the pedigrees that we have selected to illustrate the practice of different breeders, many instances may be found of high-bred animals that are good breeders. From the extent to which in-and-in breeding was carried in the case of Mr. Robert Colling's cow Clarissa, the pedigree in Diagram 8 is of particular interest. Clarissa was calved in 1814, and produced calves in 1817, 1818, and 1819, as shown by the "Herd-Book." Her daughter Restless, got by Lancaster (360), who was more than a half-brother to Clarissa, was breeding at six and seven years of age. Lancaster, the highest-priced animal at R. Colling's sale in 1818 (six hundred and twenty-one guineas), proved a valuable sire, and his name is to be found in many of the best pedigrees of the present day.

The Short-Horn bull Grazier (1085) was closely inbred (*see* Diagram 9), and a good breeder. Grazier was bred by Mr. Wiley, of Brandsby, near York. He was used in the herds of the Earl of Carlisle, Lord Feversham, Sir John Johnstone, Sir I. Ramsdon, Mr. William Smith, and Mr. Slater. He died at Byram when fourteen years of age. Twenty-seven sixty-fourths of his blood was that of Favorite, who appears for the first time in the pedigree at the third generation back.

At the Milcote sale in England, March 28, 1860, thirty-one descendants of the cow "Charmer," including "three old cows" and several calves, sold for £2,139 18s., an average of over £69 each.

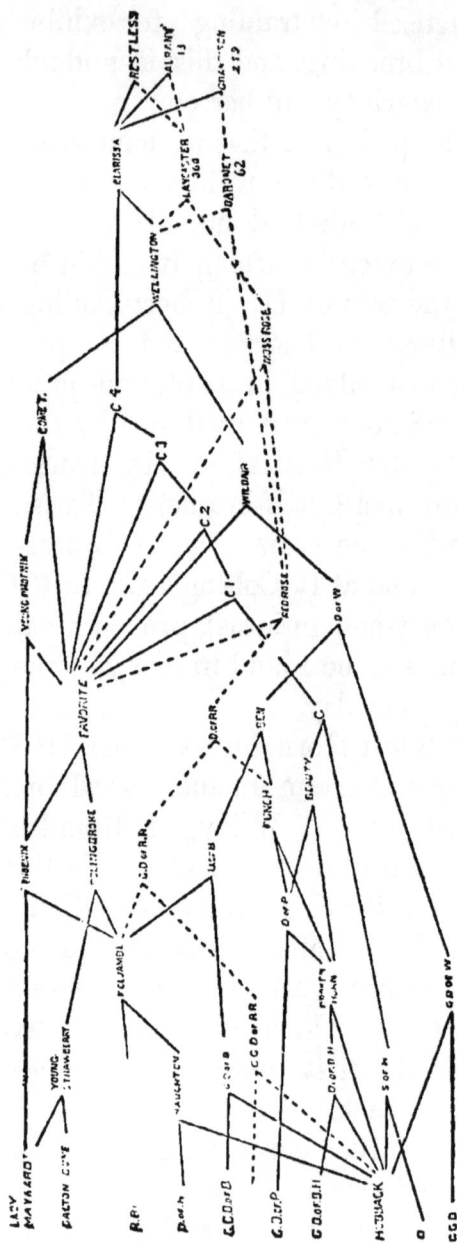

DIAGRAM 8.—PEDIGREE OF "COLLING" SHORT-HORNS.

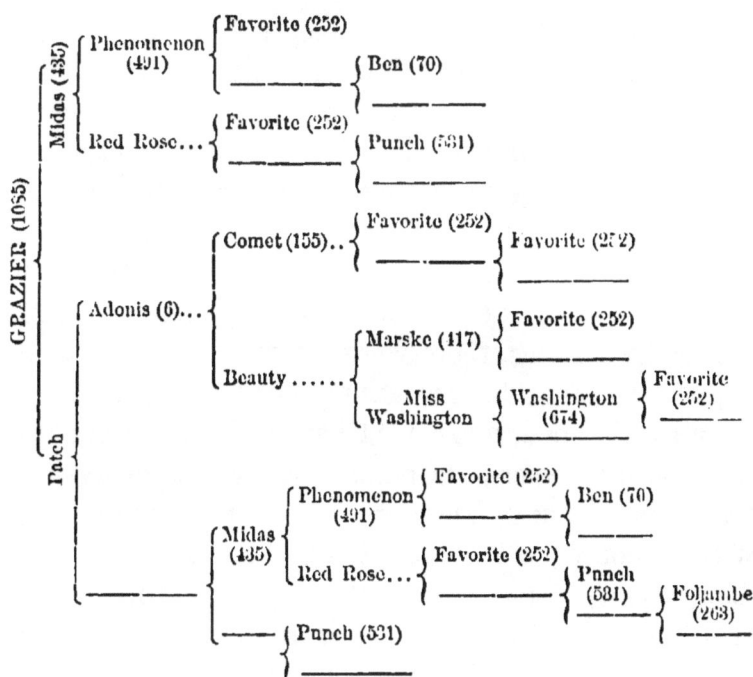

DIAGRAM 9.—PEDIGREE OF SHORT-HORN BULL GRAZIER.

Charmer " was a most extraordinary milker," and
the herd descended from her were said to be " capital
milkers, and very prolific, not having been pampered.
. . . Of the eight bulls named in the fourth generation
from which she is descended, one was ' Favorite.' She
is one-sixteenth Favorite, therefore, on that account.
But the cow to which he was then put was also descended from ' Favorite,' and so are each of the other
seven bulls and seven cows which stand on the same
level of descent with the great-great-grandam of
' Charmer.' And, in fact, it will be found, on examination, that in so far as ' Charmer's ' pedigree is
known, which it is in some instances to the sixteenth

generation, she is not one-sixteenth only, but nearly
nine-sixteenths of pure 'Favorite' blood. This arises
from 'Favorite' having been used repeatedly on cows
descended from himself. . . . In the case of 'Charmer,'
we find of her great-grandams one was the produce
of 'Favorite.' None of her progenitors in the immediately preceding generation were the produce of
that bull, but of those in the next and successive generation preceding, there were, so far as known, 2, 8,
25, 58, 101, and 99, respectively got by him. . . .

"In the pedigree of 'Charmer' we repeatedly
meet with 'Comet;' 'Comet' was by 'Favorite,' and
his dam, 'Young Phœnix,' was also by 'Favorite,'
with 'George;' 'George' was by 'Favorite,' and his
dam, 'Lady Grace,' was also by 'Favorite,' with
'Chilton;' 'Chilton' was by 'Favorite,' and his dam
also was by 'Favorite,' with 'Minor;' 'Minor' was
by 'Favorite,' and his dam also was by 'Favorite,'
with 'Peeress;' she was by 'Favorite,' and her dam
also was by 'Favorite,' with 'Bright Eyes;' she was
by 'Favorite,' and her dam also by 'Favorite,' with
'Strawberry;' she was by 'Favorite,' and her dam
by 'Favorite;' 'Dandy' and 'Moss Rose' among the
cows, and 'North Star' among the bulls, are also of
similar descent." [1]

An examination of the pedigree will show that
Charmer traces four hundred and eight lines of descent to Favorite, and that bulls descended from half
brother and sister were used eighteen times, as follows: Midas five, Barmpton six, Young Lancaster

[1] *Gardener's Chronicle and Agricultural Gazette*, 1860, pp. 270, 271,
279, 294. *See* also Goodale's "Principles of Breeding," p. 97.

three, Sultan two, and Pope and Northampton once each. Twenty-four times animals appear whose dam was daughter of their sire. Seventeen of the animals sold were got by the bull "Mameluke," who was "also full of 'Favorite' blood."

In this family we find a good illustration of the principle to which we have already called attention—that "good-milking" qualities and "good-breeding" qualities are usually associated—and this may safely be attributed to the animals "not having been pampered," and the inheritance of ancestral characters.

"M. Beaudowin gives the particulars of a flock of merinos, bred in-and-in for twenty-two years without a single cross, and with perfectly successful results, there being no sign of decreased fertility, and the breed in other respects having been improved."[1]

It has been claimed that the statistics obtained in asylums for the insane, the idiotic, and the deaf, dumb, and blind, show that consanguineous marriages are a fruitful source of this class of defects. As statistics of this kind are not readily accessible to the general reader, the following synopsis of the reports relating to this subject will be of interest in this connection.

In the examination of this class of facts, however, it must be remembered that such effects, when proved to exist, would be more marked in the human family, where there is a high development of the intellectual faculties, than among the lower animals that possess an organization that is more symmetrically balanced.

"Dr. Chazarain, a young physician of Bordeaux,

[1] *Comptes-Rendus,* August 5, 1862; according to "Transactions of New York State Medical Society," 1869, p. 111.

has written a very able thesis on the same subject, which contains numerous observations on the influence of consanguinity on deaf-dumbness. It appears that in the Deaf and Dumb Institution at Bordeaux, of thirty-nine boys deaf and dumb, six were the offspring of such marriages; and of these six, one boy had two brothers deaf and dumb, and one boy had three brothers deaf and dumb, making a total of eleven.

"Of twenty-seven girls, in the same institution, nine were the issue of such marriages; and of this number, six had between them seven brothers and sisters similarly affected, making a total of sixteen; and very lately (1860), M. Devay, Professor of Clinical Medicine at Lyons, has brought the same subject before the notice of the Imperial Academy of Sciences, in that city; for, to so great an extent has the evil prevailed, that in one of the departments of France (Arièges), the clergy have endeavored to check the frequency of such marriages, and have appealed to the authorities at Montpellier to aid them in so doing." [1]

"In a very able paper 'On Marriages of Consanguinity and Deaf-Dumbness,' which is generally supposed to be one of the most constant defects resulting from such marriages, M. Boudin informs us that 'deaf-mutes are the issue of consanguineous marriages in the proportion of twenty-eight per cent. at the Paris Imperial Institution, twenty-five per cent. at Lyons, and thirty per cent. at Bordeaux;' and that as regards the

---

[1] Sedgwick, in the *British and Foreign Medico-Chirurgical Review*, July, 1861, p. 143.

Jews in Berlin, 27 in 10,000 are deaf-mutes, while the proportion is only 6 in 10,000 among the Christian population in that city; and apparently, therefore, with great justice, he concludes that 'the hypothesis of the pretended harmlessness of consanguineous marriages is contradicted by the most evident and well-authenticated facts.' " [1]

Statistics have been collected, apparently showing that the cousins of persons who are deaf and dumb are particularly subject to the same defect, and it has been inferred that they furnish a satisfactory explanation of the preceding statements. As we wish to present as strong a case as possible in favor of the theory that consanguineous marriages are likely to result in defects in the offspring, the following statistics are added:

From Wilde's "Report on the Deaf and Dumb of Ireland" (as quoted by Sedgwick), it appears that "in cases of single congenital mutism, where the relations were also deaf and dumb, there were by the father's side . . . . in one instance six cousins affected; in three cases there were four cousins, and in nine cases two cousins all deaf-mutes. Where two of the family were affected with congenital deaf-dumbness, in two instances four cousins were in a similar condition; in two cases three cousins were deaf and dumb, and in four instances two cousins were thus affected.

"Where three cases of congenital deaf-dumbness occurred in the same family . . . . in five cases, two cousins were in a similar condition. When the rela-

---

[1] *British and Foreign Medico-Chirurgical Review*, July, 1863, p. 179.

tionship came by the mother's side, there were in cases of congenital mutism eight cases where three cousins were deaf and dumb, and fourteen instances where two cousins were thus affected. In the case of two mutes in the same family, we find that . . . . in one case three cousins, and in four cases two cousins, labored under the like defect . . . . and . . . . where three mutes occurred in the same family . . . . two cases presented of two cousins also deaf and dumb."

These facts, upon their face, appear to be conclusive as to the alleged influence of consanguineous marriages in producing the defects in question. In collecting them, however, a number of important points have been neglected, which seriously detract from the force they would otherwise be entitled to. Endemic conditions, we have already observed, have a decided influence in producing similar malformations, and an inherited tendency of the parents to a diseased condition of the system of another form would likewise aid in their production.

In this connection it will be well to examine the facts in relation to the heredity of deaf-mutism. Dr. Joseph Adams,[1] in noticing the statistics furnished by the Deaf and Dumb Institution of London, says: "Of one hundred and forty-eight scholars upon the foundation of this institution, one is of a family where there are five deaf and dumb (himself included); one where there are four; eleven where there are three; and nineteen where there are two.

"Of the scholars, fifty-seven are girls, and the rest

[1] "Hereditary Diseases," p. 66.

boys; *none of them of deaf and dumb parents.* The gentleman who superintends the manufactories, and who, consequently, has the best opportunity of tracing the subsequent history of his scholars, informs me that some of them are married and have children, all of whom are perfect in their organs of hearing. One instance has occurred, in which both parents were born deaf, yet their children hear."

"At the school for the deaf and dumb in Manchester (England), in 1837,[1] there were forty-eight children taken from seventeen families . . . . an average of nearly three such cases in each family. Out of these instances there appears but one in which the defect was known to exist in either parent."

The following cases are likewise quoted from Mr. Sedgwick's valuable article on " Hereditary Diseases :" " Mr. Wilde, whose observations included the whole of the deaf and dumb population of Ireland, states, that 'ninety-eight deaf and dumb persons — sixty males, and thirty-eight females—were married. In eighty-six instances—fifty-four males, and thirty-two females—only one parent was deaf and dumb: from the marriage of these, two hundred and three children resulted, among whom there was but one instance of mutism, a male, in the county of Limerick. Six instances have been recorded of the intermarriage of deaf and dumb persons : their offspring amounted to thirteen, of whom only one, a female, in the city of Dublin, was deaf and dumb.' "

" Lastly, in the thirty-fifth annual report of the

[1] "Notes and Reflections," by Sir Henry Holland, as quoted by Sedgwick, *loc. cit.*, p. 142.

asylum in Hartford (United States), we find that one
hundred and three of the deaf and dumb had been,
or are now, married. In forty-one of these marriages,
both parties were deaf and dumb; in twenty-three,
one could speak or hear. Of these one hundred and
three, thirty-one had not become parents, but the remaining seventy-two were parents of one hundred and
two children, of whom ninety-eight could hear and
speak, and four only were deaf and dumb. One of
the four was the only child of his parents, both of
whom were congenitally deaf. Besides the parents,
the .paternal grandfather, a sister of the father, and
two sons of this sister, were deaf and dumb. In the
other family, that of three children, the father lost
his hearing by disease at two years of age, and had
no known relative deaf and dumb. The mother was
born deaf, and had a deaf and dumb brother." [1]

In commenting upon the fallacy of inferences
drawn from the preceding statistics, and particularly
on those presented by M. Boudin in his paper on
consanguineous marriages above quoted, Mr. Sedgwick says : " In the first place, with regard to the supposed frequency of consanguineous marriages among
the Jews, M. Isidor, the Grand Rabbi of Paris, states
that such marriages are far less frequent than is generally believed ; and, moreover, if the inference drawn
from the great prevalence of deaf-mutism among the
Jews of Berlin were correct, the statistics would be
found to coincide with those of deaf-mutism among the
Jews elsewhere, but such is not the case, for, although

[1] *British and Foreign Medico-Chirurgical Review,* July, 1861, p.
143.

the number of Jews in Paris is estimated at twenty
five thousand, only four of them are deaf-mutes.
Again, with respect to the greater frequency of deaf-
mutism in other races among the offspring of those
who are allied, compared with those who are aliens
by blood, although the facts adduced by M. Boudin
and other writers are undoubtedly correct, yet the inference that has been drawn from them is in like manner probably erroneous, for all, or nearly all, the illustrations of deaf-mutism in these cases of consanguinity,
have occurred in circumscribed localities, where deaf-
mutism, independent of consanguinity, is more common than elsewhere. Mr. W. R. Scott, of the Deaf
and Dumb Institution at Exeter, has lately called attention to the fact that deaf-mutism occurs in much
larger proportion in secluded and rural populations
than in urban and manufacturing districts; in the
union of Crediton, in Devonshire, one in 1,143 of the
population is a deaf-mute, and in the Scilly Islands
this is still more remarkably shown by the fact that
there are no less than six deaf-mutes in a population
of 2,677, or one in 446. But perhaps the strongest
argument against the unqualified admission in these
cases of consanguinity as the *fons et origo mali*, is
the fact that deaf-mutism cannot as a rule be directly
transmitted to the offspring even in those cases in
which both the parents are deaf-mutes; for it is chiefly
by means of breeding-in that peculiarities of structure
among the lower animals are perpetuated; and their
hereditary transmission is effected with so much certainty and facility that it would be difficult, in the
present day, to say what amount of abnormal develop-

ment may not by this system be established as a permanent variety.

" It is therefore evident that consanguinity alone cannot be accepted as the cause of deaf-mutism, nor consequently as the sole cause of any other diseases or defects which have from time to time been ascribed to it." [1]

In 1858, Dr. Bemis, of Kentucky, made a report [2] to the American Medical Association on "Marriages of Consanguinity," in which he gives an imperfect history of eight hundred and seventy-three instances of such marriages.

The conclusions of the author are briefly stated in the paper as follows : " I feel satisfied, however, that my researches give me authority to assume that over ten per cent. of the deaf and dumb, and over five per cent. of the blind, and near fifteen per cent. of the idiotic in our State institutions for subjects of those defects, and throughout the country at large, are the offspring of kindred parents, or of parents themselves the descendants of blood intermarriage."

It will be observed that Dr. Bemis does not assume that the relationship of the parents is the cause of the defects of their children, and there is nothing in the report to warrant such a conclusion. The facts presented in the report are of particular interest, as they furnish a good illustration of the difficulty of obtaining exact statistical information on subjects of this kind. One source of fallacy, arising from the manner

---

[1] *British and Foreign Medico-Chirurgical Review*, July, 1863, pp. 179, 180.

[2] "Transactions of the American Medical Association," vol. ii., p. 319.

in which the statistics were collected, is noticed by Dr. Bemis, who says: "But while indorsing the truthfulness of these statistics, it is my duty to state that those which relate to marriages of consanguinity should probably not be received as a completely true representation of the results of such marriages; some modification of the mean of results might occur if the statistics of *all* instances of in-and-in marrying, in the Union, for example, could be comprised in one report. It is natural for contributors to overlook many of the more fortunate results of family intermarriage, and furnish those followed by defective offspring or sterility. The mere existence of either of these conditions would prompt inquiry, while the favorable cases might pass unnoticed."

It is well known that the defects in question may be produced by a variety of causes, but their presence or absence in the cases reported is not noticed.

When a predisposition to these or similar defects exists in a family, the intermarriage of its members would, without doubt, result in their repetition in the offspring by direct transmission, and the influence of the relationship of the parents could not, as a matter of course, be determined.

In nearly all the cases collected by Dr. Bemis, the history of the ancestors of the parents is not given, and there is therefore nothing to show that the defective children in the cases reported have not been afflicted by a direct inheritance of their deformity. In one hundred and eighty-one of the cases, one or both parents are reported as delicate in constitution, addicted to bad habits, or suffering from disease, and in over one hundred cases no report is given as to the health or habits of the parents.

The mere fact that the parents of these defective children were related, throws no light upon the cause of their infirmities, which can only be determined by a knowledge of details that the report does not furnish.

Dr. Robert Newman, of New York, as chairman of a committee appointed for that purpose, made a " Report on the Result of Consanguineous Marriages" to the New York State Medical Society, from which we make the following extracts, showing the opinions of able men who have given the subject a careful examination, and a summary of the results of the inquiries made by the committee.

Dr. Gilbert Child says, " The marriages of blood relations have no tendency, *per se*, to produce degeneration of race."

Prof. S. II. Dickson, of Philadelphia, in his lectures on "Scrofulosis and Tuberculosis," makes the following statement: "Several writers on both sides of the Atlantic—on this side Prof. Bemis—ascribe much of tuberculosis and scrofulosis to the marrying of relatives—physical incest, as it is called. I think the truth can be put in a nutshell. I suggest it to you, there is a great deal of exaggeration on this subject, yet there is much reason for the belief that the intermarriage of relatives is dangerous to the offspring, not on account of their mere consanguinity, but because they are likely to have the same predisposition to scrofula, if that predisposition exists in that family. . . . Therefore we come to the conclusion that it is not an essential result of marriage of consanguinity that there should be scrofulous or other degeneracy." Dr. Edward Jarvis, the distinguished statistician, in a letter to Dr. Newman, says: "Cousins, descendants from a common ancestry, have a common heritage—of good, of evil, of power, and weakness ; and, if these join in marriage, their issue have a double chance of inheriting whatever qualities they may both possess. If, then, both parents, although cousins, are perfect in constitution and health, and have nothing to transmit but power, then their children have a double security against constitutional imperfection, and a double war-rantee of inherited capacity and strength. The converse is also true with cousins who have imperfections and liabilities in common. If they marry, they provide a double chance of the repetition of the same weaknesses and susceptibilities in their offspring. . . . In this view of the matter, the objection to consanguineous marriages lies not in the bare fact of their relationship, but in the fear of their having similar vitiations of constitution."

Dr. Newman gives the details of thirty-two instances of consanguineous marriages, in different localities. The result, as far as reported, was one hundred and twenty-seven children, or nearly an average of four to each marriage, and there were instances of eight, eleven, twelve, and even fourteen children in a family ; while but one marriage proved unproductive, but in this case both parties were affected with disease. Of the one hundred and twenty-seven children, but fourteen died under two years of age, which is eleven per cent. ; while in Michigan, in 1870, according to the reports of the State Board of Health, the mortality of children under two years was nineteen and one-half per cent., and in the metropolitan district of New York, in 1868, it was thirty-eight per cent.

Dr. Newman says: "With regard to scrofulous children we observe as follows: Either parent, or both, we find scrofulous or tuberculous in six cases, Nos. 3, 5, 16, 18, 20, and 21, the offspring of which were, so far, fifteen children, of which four died young, a common percentage; in reference to health, we find five scrofulous and ten healthy, therefore we have from partly unhealthy parents two-thirds healthy children."

"In regard to healthy or unhealthy organization, we find of these one hundred and twenty-seven children deviating from a perfect state, as follows." Five scrofulous, above mentioned; one case of epilepsy and one of amaurosis in the same family, with twelve other children not thus affected; one case of two children in a family "having only two phalangeal bones in the index-finger, otherwise they are reported as healthy and intelligent;" and two deaf-mutes in one family. The cases of the deaf-mutes and the child said to be "simple" occurred at Panama under circumstances not favorable to healthy development. In the same report,[1] Dr. Newman says: "We cannot but notice here a fact communicated by Dr. H. Knapp, late professor in the University of Heidelberg, which we add to the statistics: In Nassau (Germany), only three families established the village of Dauborn, and kept entirely isolated. Their children, therefore, intermar-

[1] "Transactions of the New York State Medical Society," 1869, pp. 109-130.

ried; and at present the village has fifteen hundred inhabitants, who are of strong constitution, and are active, sprightly, intelligent, and healthy. Our informant had this place directly under his observation, and says he neither saw deformity nor insanity, and only one case of deaf-mutism; in fact, the entire race was robust and healthful."

Dr. T. A. McGraw, who has written an interesting article on this subject, says: "There can be no doubt that close and continual interbreeding has taken place, time and again, without any evident injurious consequences among simple and uncultivated communities. Notable examples are the Pitcairn Island settlement, formed from the close in-and-in breeding of the progeny of four mutineers from the ship Bounty, and nine native women; the small community of fishermen near Brighton, England; the numerous small and isolated villages of Iceland; and the Basque and Bas-Breton settlements among the Pyrenees. . . . We must admit, from overwhelming evidence, that under such circumstances as the settlements just mentioned afforded, consanguinity among married people does not necessarily cause evil results to the progeny. If it is asked how it would be with men of more civilized habits, we are unfortunately obliged to confess that there are no statistics whatever on the subject which can give us any exact and trustworthy information."[1]

Dr. Mitchell, of the Edinburgh College of Physicians, says of idiocy and its relations to marriages of consanguinity, that in more than sixty per cent. of the

[1] Dr. T. A. McGraw's "Address on Heredity and Marriage, pp. 12, 13.

cases occurring in the British Isles the condition is acquired and not congenital.[1]

Of fifteen hundred and fifty-seven patients in the insane asylums of Paris, Auguste Voisin found none that were the result of consanguinity.[2]

The facts that have thus far been collected in regard to this subject seem to warrant the conclusion that close breeding, in itself considered, is not injurious; but, as it tends to fix and perpetuate the constitutional defects that have been produced by other well-known agencies, it should not be practised by careless or inexperienced persons, who do not make a judicious selection of their breeding-stock, as they are likely to obtain, through its influence, the most unsatisfactory results.

The most obvious objection to close breeding—and it is perhaps the only one of importance—is the difficulty of selecting animals that are free from constitutional defects, and the danger arising from the tendency of such defects to become dominant in the offspring.

It must, however, be admitted that it is an important means of improvement when judiciously practised, and that it constitutes the only known method of securing an accumulation of the slight variations, in a particular direction, that it may be desirable to retain and perpetuate.

The greatest improvement in the form and quali-

---

[1] *Popular Science Monthly*, June, 1872, p. 250.

[2] London *Lancet*, quoted in the *Popular Science Monthly*, December, 1873, p. 179; and in the *Pacific Medical and Surgical Journal*, February, 1877, p. 408.

ties of animals can only be made by those who possess the requisite knowledge and skill to enable them to blend and perpetuate all desirable variations, through a system of rigorous selection and close breeding, without impairing the constitution by an accumulation of undesirable characters.

CHAPTER X.

CROSS-BREEDING, strictly speaking, is the pairing of animals belonging to distinct breeds, and, in this limited sense, it may be considered the opposite of in and-in breeding.

The terms "crossing," "making a cross," "out-breeding," and "cross-breeding," are, however, frequently used to indicate the mixture of the blood of different families belonging to the same breed. As the principle involved in both of these methods is the same, and the loose use of these terms is not likely to lead to any serious confusion in their application, we need not attempt to assign them a more definite meaning.

The advantages of cross-breeding have been strongly urged by a large number of writers, and in many instances it is undoubtedly the best possible practice. The improvements that have been effected by crossing, in particular cases, have been, however, without sufficient reason attributed to some direct influence arising from the process itself. Some of the best authorities on this subject have evidently been misled in their attempted explanation of the advantages of the system, by placing too much confidence in the theory that the male has the greatest influence in determining the form and general qualities of the offspring.[1]

Before attempting to ascertain the effects of cross-breeding, in itself considered, it will be well to examine some of the cases in which it has been successfully practised, with especial reference to the conditions that made it desirable.

Cross-breeding has, perhaps, been practised to a greater extent with sheep than with any other class of animals, and among them we find the best examples of well-established cross-breeds. In Hampshire the old horned breed of sheep, with its " large bones," " high withers," and sharp spine, was crossed with rams of the improved Southdown, until its original defective characters almost entirely disappeared.

The improved Hampshire are characterized by the absence of horns, "a broader back, rounder barrel, shorter legs, and superior quality altogether."[2]

Some of the leading breeders of the Hampshires were not content to rely upon the general superiority of the improved Southdowns in improving their flocks, but obtained the best rams that could be found in the flocks of the most celebrated breeders. Mr. William Humphrey, a noted Hampshire breeder, it is said, sent to Jonas Webb for one of " his best sheep," and Mr. Spooner attributes his success to a great extent " in seeking his improvements from such a renowned flock."[3]

[1] *Quarterly Journal of Agriculture*, vol. i., p. 34 ; *Journal of the Royal Agricultural Society*, vol. xx., pp. 294–310.
[2] Ibid., p. 300. [3] Ibid., pp. 305–312.

In Wiltshire, where the same old horned stock originally prevailed, a different system was practised —the improved Southdown gradually took the place of the old breed, which soon disappeared. The imported Southdown ewes were after a time crossed with improved Hampshire rams, that already had a large proportion of Southdown blood, for the purpose of giving an increase in size.[1]

The Morfe Common sheep of Shropshire were a small, fine-wooled race, accustomed to short pastures and scanty fare. The improved Shropshire, the result of a cross of the old race with the Cotswold, Leicester, or Southdown, is larger, more compact, fattens more rapidly, and in general qualities is better adapted to an improved system of husbandry.

The new Oxfordshire breed, which is highly prized in many localities, was obtained by crossing Cotswold rams on Hampshire or West Country Down sheep.[2]

" There are few districts in England in which some advantage has not been derived from the cross-breeding of sheep. Even the little *mountain-sheep* of Wales has been greatly improved by the Cheviot ram, a *larger, superior*, but still a mountain-sheep.

" At the same time the Cheviots themselves have been improved for the butcher by crosses with the Leicester, the Cotswold, and the Down. Their progeny have been increased in size, and fatten more readily." [3]

---

[1] *Journal of the Royal Agricultural Society*, vol. xx., p. 303.

[2] Ibid., p. 308.

[3] Spooner on " Cross-Breeding." *Journal of the Royal Agricultural Society*, vol. xx., p. 309.

In all of these cases the object has been to improve an inferior breed by ingrafting upon it the superior characteristics of another. The improvement has been produced, not from the fact that a male of another breed has been used, but from the higher breeding and superior qualities of the males thus selected. The superior male is found to be prepotent when coupled with females of inferior quality.

The experience of breeders in making a cross of Cheviot rams upon the ewes of the Black-Faced Heath breed will furnish some important suggestions in regard to the real causes of improvement.

"In this cross," says the intelligent Scotch shepherd, William Hogg, "the independent habits of the mountain-flocks were lost, and a mongrel progeny, of a clumsy figure, occupied the lowest and warmest of the pastures." The cross-bred animals, although retaining largely the characteristics of the original breed, were not able to withstand the "hardships and cold of winter," and they required better care and better pastures than the old race had been accustomed to.

"Another truth which the process of changing a numerous stock has disclosed is, that, in the produce of the first crop, and for several successive issues, the figure, wool, and other qualities of the Cheviot ram, are most conspicuous in the smallest and feeblest of the progeny; while the properties of the mountain-breed are more fully exhibited in the strongest and most robust of the lambs. This misled many of the storemasters. They did not consider that there was as much Cheviot blood in the coarsest (as they were pleased to call them) as in the finest; though not so

clearly exhibited in its external qualities. This induced them to throw aside the best of the lambs and
select those to breed from which had apparently most
of the Cheviot figure. This was an additional disadvantage; for, as it prevailed wherever the experiment was tried, the mountain-flocks in general were
smaller and feebler than ever they were known to
have been; and were, consequently, more vulnerable
to bad seasons, a course of which happened to accompany the change." [1]

The stability of the characteristics of the old
mountain - breed was shown in the readiness with
which the cross-bred animals were "bred back" to
the original type, and the frequent appearance of the
old characters by atavic descent after an effort for
twenty-five years to establish the peculiarities of the
Cheviot.

"The black-faced sheep," says Youatt, "seemed
obstinately to resist the influence of foreign crosses.
The Leicester, and even the Cheviot blood, added
little to the value either of the fleece or the carcass,
while they materially lessened the hardihood of the
sheep." [2]

Sir John Sinclair also observes that "the Dishley
breed is perhaps the best ever reared for a rich, arable
district; but the least tincture of this blood is destruc-
tive to the mountain-sheep, as it makes them incapable
of withstanding the least scarcity of food." [3]

The Cheviots, although a mountain-breed adapted

[1] *Quarterly Journal of Agriculture*, vol. i., p. 178.
[2] Youatt on "Sheep," p. 325.
[3] As quoted by Youatt, *loc. cit.*, p. 325.

to moderate elevations and better pastures than the black-faces required, were decidedly improved by a cross of the improved Leicester,[1] the conditions in which they were placed admitting of a class of animals of better-feeding quality.

The cross of a superior breed on one that is inferior cannot, then, succeed in producing improvement without being accompanied by better management and more liberal feeding. After the times described by Hogg and Youatt, the Cheviots were extensively introduced in the Highlands of Scotland, and their success is an evidence of an improved condition of agriculture. In the cross-breeding of cattle and horses the same influences have determined, to a greater or less extent, the success or failure of the practice.

The advocates of a system of cross-breeding, almost without exception, insist upon the importance of making use of males of superior character in all essential qualities. " Having duly recognized the claims of thorough-bred horses of the first and second class," says Mr. Spooner, " we can only advise, with regard to the third and inferior classes, that their services be altogether dispensed with, their place being taken by three-fourths or half-bred stallions, possessing bone, substance, and good hunting qualifications."[2]

He prefers "the services of a first-class, thorough-bred stallion . . . . on the rare occasions when they are offered "—but, when they cannot be procured, a part-bred stallion is to be selected, provided he is better in

---

[1] Youatt, *loc. cit.*, p. 335.

[2] *Journal of the Royal Agricultural Society*, 1865, p. 165.

the points it is desirable to perpetuate than the pure-bred stallions that are within reach.

The cross of the pure-bred stallion upon mares of mixed blood is not, then, to be recommended if the stallion is inferior in its characters; and the general rule of breeding, to which there are no exceptions, that the best males it is possible to obtain should only be used, becomes the guide in practice.

In an article " On Cross-Breeding Cattle," Mr. Murray says: " The importance of using, even for cross-breeding, none but first-class bulls, can hardly be sufficiently insisted upon. Indeed, the marked success which has attended the use of Short-Horn bulls may be attributed not less to their established position than to the intrinsic merits of the race;"[1] and he adds, " We are fully convinced that, even for the purpose of cross-breeding, the purer the blood on the paternal side, the more clearly will excellence be stamped on the progeny." The same writer attributes the failure, in cases of unsuccessful crossing, to the use of inferior bulls that were not able to impress any superior qualities upon their offspring.[2]

Short-Horn bulls have, undoubtedly, been more extensively used in crossing other breeds than any others; but, when the cross has been successful, it can only be attributed to the higher breeding and superiority of the typical characters of such bulls, which enabled them to stamp their peculiarities upon the carelessly-bred stock they were selected to improve.[3]

---

[1] *Journal of the Royal Agricultural Society*, 1866, p. 53.

[2] *Loc. cit.*, pp. 53, 54.

[3] *Journal of the Royal Agricultural Society*, vol. xxiii., p. 351.

The male must not only possess superior merit in his general characteristics, but he must have the essential quality of prepotency in transmitting them.

One difficulty in the way of ingrafting the characteristics of the Cheviot sheep upon the Black-faced mountain-breed, to which we have referred, arose from the uniform typical characters and consequent prepotency of the race it was proposed to change.

In crossing English rams upon the old-established breeds of France, the same difficulties were experienced, viz., the prepotency of the French stock, and the fact that the English breeds were not adapted to the climate and system of management they were subjected to in their adopted country.

M. Malingie-Nouel, director of the agricultural school of La Charmoise, has given his experience in establishing the Charmoise breed, from which we make the following quotations : [1]

" When an English ram of whatever breed is put to a French ewe, in which term I include the mongrel merinos, the lambs present the following results : Most of them resemble the mother more than the father; some show no trace of the father; a very few represent equally the features of both. Encouraged by the beauty of these last, one preserves carefully the ewe-lambs among them, and when they are old enough puts them to an English ram.

" The products of the second cross, having seventy-five per cent. of English blood, are generally more like the father than the mother, resembling him in

---

[1] Translated by Mr. Pusey, in the *Journal of the Royal Agricultural Society*, vol. xiv., p. 214.

shape and features. The fleece also has an English character.

"The lambs thrive, wear a beautiful appearance, and complete the joy of the breeder. He thinks that he has achieved a new cross-breed insuring great improvement, and requiring thenceforth only careful selection to perpetuate by propagation among themselves the qualities which he has in view. But he has reckoned without his host. For no sooner are the lambs weaned, than their strength, their vigor, and their beauty, begin to decay as the heat of our summer increases. Instead of growing, they seem to dwindle; their square shapes shrink; they become stunted; and, on the threshold of life, put on the livery of old age.

"A violent cold in the head completes their exhaustion. This is accompanied with a copious flow of slimy mucus from the nostrils, constant sneezing, and sometimes cough. At last the constitution gives way, or, if the animal lasts till autumn, the malady indeed ceases, but he remains stunted for life.

"The time lost was the time of growth, and cannot be recovered, for Nature never goes backward. Henceforth he looks like a foreigner escaped from the mortal influence of an inhospitable climate, and remains inferior even to our native sheep, which at least have health and hardiness in their favor. The experiment has sometimes been tried with English rams in a third generation, and the symptoms above described have arisen even more strongly in proportion to the stronger admixture of English blood."

After pointing out some differences observed in

the prepotency of several English breeds, M. Malingie-Nouel says: "If you put a Leicester ram, a *mixed* New Kent,[1] or a Southdown that is *not pure*, to a pure ewe of any French race, very little English character is impressed on the offspring, never less than when the ewe is a pure merino. In this last case, it often happens that you can see no difference between lambs that are Leicester-merinos, Kent-merinos, or Southdown-merinos, and another lamb of the same age which is pure merino. In compensation, however, for this feeble influence of the English sire, the lambs of such first crosses have no more difficulty than French lambs in getting over the first summer. If, on the contrary, the same ewes are put to *very pure* rams of the Southdown or New Kent breed, the English character is more marked than in the former cases.

"In both cases the offspring is reared; for lambs in which the English blood does not exceed one-half seem to be reared as easily as pure French lambs. But, then, since little improvement is obtained, one is tempted to give a new dose of English blood—to put the Anglo-French ewes to English rams—whereupon the disasters described are sure to follow."[2]

After discussing the causes of the state of facts above referred to, M. Malingie-Nouel proceeds as fol-

---

[1] The New Kent breed was established by Richard Goord, from "nine ewes and one ram" of the Romney Marsh breed, and a few rams obtained of Mr. Wall. They were deeply in-bred, and like the Southdowns were improved without crossing (*Journal of the Royal Agricultural Society*, vol. vi., p. 263).

[2] *Loc. cit.*, pp. 217, 218.

lows : "It appeared, then, that in order to untie the Gordian knot whose threads I have traced, inasmuch as one could not increase the purity and antiquity of the blood of the rams, one must diminish the resisting power, namely, the purity and antiquity of the ewes. With a view to this new experiment, one must procure English rams of the purest and most ancient race, and unite with them French ewes of modern breeds, or rather of mixed blood forming no distinct breed at all. It is easier than one might have supposed to combine these conditions.

"On the one hand, I selected some of the finest rams of the New Kent breed, regenerated by Goord. On the other hand, we find in France many border countries lying between distinct breeds, in which districts it is easy to find flocks participating in the two neighboring races.

" Thus, on the borders of Berry and La Sologne, one meets with flocks originally sprung from a mixture of the two distinct races that are established in those two provinces. Among these, then, I chose such animals as seemed least defective, approaching, in fact, the nearest to, or rather departing the least from, the form which I wished ultimately to produce. These I united with animals of another mixed breed, picking out the best I could find on the borders of La Beauce and Touraine, which blended the Tourangelle and native merino blood of those other two districts. From this mixture was obtained an offspring combining the four races of Berry, Sologne, Touraine, and merino, without decided character, . . . but possessing the advantage of being used to our climate and management, and bringing to bear on the new breed to be formed an influence almost annihilated by the multiplicity of its component elements. Now, what happens when one puts such mixed-blood ewes to a pure New Kent ram?

"One obtains a lamb containing fifty-hundredths of the purest and most ancient English blood, with twelve and a half hundredths of four different French races, which are individually lost in the preponderance of English blood, and disappear almost entirely, leaving the improving type in the ascendant. The influence, in fact, of this type was so decided and so predominant that all the lambs produced strikingly resembled each other, and even Englishmen took them for animals of their own country.

"But what was still more decisive, when these young ewes and rams were put together, they produced lambs closely resembling themselves, without any marked return to the features of the old French races from which the grandmother ewes were derived. Some slight traces only might perhaps be detected here and there by an experienced eye. Even these, however, soon disappeared, such animals as showed them being carefully weeded out of the breeding-flock."[1]

Such was the origin of the Charmoise breed of sheep.[2]

M. Girou "supposed that he would more speedily obtain fine wool by crossing Roussillon sheep with

[1] *Loc. cit.*, pp. 220, 221.

[2] For a full description of this valuable breed, *see* "Encyclopédie Pratique de l'Agriculteur," tome x., p. 582.

Merino rams, than by uniting the Aveyron breed with the same rams, but he was disappointed.

"The Roussillon race, being no doubt more ancient and possessed of greater potency (*force motrice*) than the Aveyron race, offered greater resistance than the latter to the influence of the Merinos; and, after twenty-five years of successive crossing, the primitive characters of the Roussillons still appeared, while the crosses of the Aveyron race, after the same length of time, could not be distinguished from the Spanish sheep. It thus appears that characters long established, and thoroughly incorporated with the constitution by transmission through many successive generations, give to a race or breed a certain fixity of type—something of the persistency and individuality of a species, by which it is enabled to resist, for a length of time, fusion with another race, and continue to reproduce its leading characteristics."[1]

It has been said that "the persons who chiefly resort to crossing are those who have, up to the present time, kept but a very inferior description of stock,"[2] and this is, without doubt, the reason why cross-breeding has been found to be, in such cases, an important means of improvement.

In all cases in which cross-breeding has been successfully practised, the object in view has been precisely the same, and the reasons that have led to it are identical with those that have induced the improvers of the pure breeds to resort to the opposite system of in-and-in breeding.

---

[1] *Journal of the Highland Agricultural Society*, 1857-'59, p. 29.
[2] *Journal of the Royal Agricultural Society*, vol. xxiii., p. 352.

In both cases the practice has been to select the best male it was possible to secure, with the purpose of impressing his superior characteristics upon the less-favored individuals of the flock or herd.

The rule laid down by George Culley, in his "Observations on Live-Stock," has apparently been followed by the advocates of both of the so-called "systems of breeding," which are in reality but parts of the one true method. He says: "It is certainly from the *best males* and *females* that the best breeds can be obtained or preserved; to breed in this manner is undoubtedly right, so long as *better males* can be met with, not only among our neighbors, but also among the most improved breeds in any part of the island, or from any part of the world, provided the expense does not exceed the proposed advantage. And when you can no longer, at home or abroad, find *better males* than your own, then, by all means, breed from them—whether horses, neat-cattle, sheep, etc.—for the same rule holds good through every species of domestic animals; but upon no account attempt to breed or cross from worse than your own, for that would be acting in contradiction to common-sense, experience, and that well-established rule, that 'best only can beget best.'"[1]

If it should be admitted that the pure breeds were better than cross-bred animals, it would be impossible for every one to obtain them in sufficient numbers to stock the farm exclusively with them; but, as well-bred males can readily be procured, the greatest improvement in the mass of our farm-stock must be made by a system of judicious crossing.

[1] "Observations on Live-Stock," fourth edition, p. 12.

The breeding of " grades," which is so largely practised in all parts of this country, furnishes a good example of the advantages of cross-breeding. What are called " natives " here, are animals of mixed blood without any fixed characters, and they are therefore more readily influenced by a cross of superior blood than the local unimproved native breeds of Europe, that have more definite characteristics.

Earl Spencer has remarked [1] that " the worse bred the female is," the greater the influence of a well-bred male upon the offspring, and this accords with the observations of practical men generally.

The originator of the Charmoise breed of sheep, as we have seen, developed the prepotency of the English rams used, by mixing the blood of the ewes of several native races, and thus destroying in them the fixed characters that had previously prevented the predominance of the desired characteristics of the English breeds.

As the dominant peculiarities of the pure-bred animal are developed by a system of rigorous selection and in-breeding in a certain definite direction, they will also as readily disappear and become latent, if the opposite practice of cross-breeding is resorted to, and this is one of the most uniform effects of this method of breeding.

If a cross of two distinct breeds is effected by the selection of animals of equal power in the transmission of their peculiar characteristics, the tendency is to make dominant the original characters that the breeds had in common, and to obscure the special

[1] *Journal of the Royal Agricultural Society,* vol. i., p. 22.

characters that constituted their distinguishing characteristics.

The greater the contrast presented in the two breeds, and the greater the specialization of their qualities through the development of artificial characters, the stronger the tendency to obscure the best characters of each, and restore the original type from which they had been developed.

In such cases the offspring would in all probability prove to be inferior in quality, from the inheritance of the defects of both parents, without retaining the most desirable characters of either.

All the best authorities on cross-breeding agree that it should not be practised without a definite object, on account of this tendency to the development of undesirable variations.

Mr. Dickson says: "I object to promiscuous crossing as much as any man. It is to this injudicious system that may be traced the existence of so many miserable breeds of cattle in the country."[1]

"We may start, then," says Mr. Spooner, "with this principle, that to cross for crossing sake is decidedly *wrong;* that, unless some specific purpose is sought for by crossing, it is far better to cultivate a pure breed."[2]

Prof. Tanner remarks that, "in the case of purebred animals, there should be no opposing influence to weaken the hereditary tendencies of the offspring, but on the other hand a concurrent and sympathetic nature, so that the hereditary character may be con-

[1] *Quarterly Journal of Agriculture,* vol. vii., p. 508.
[2] *Journal of the Royal Agricultural Society,* vol. xx., p. 298.

firmed and strengthened. Anything like a cross should be most jealously guarded against, as introducing a conflict of influences which impairs the character of the race."[1]

Mr. Wright, in discussing this subject, says, "There is, for instance, a well-known strain of Buff Cochins, of marked excellence in every point, but which has a strong tendency to breed a white feather in the cock's tail.

"Now, it is perfectly possible, by a judicious cross from some other strain, and careful selection afterward, to get rid of this objectionable feature; and we will suppose an individual yard in which this has been so far accomplished that in only a small proportion does the hated white feather appear. This desired result, with a little care, will now be easily maintained while such a yard is bred to itself, or with any other not too far removed from it in blood; but if crossed from a strain *thoroughly* distinct and alien, or what poultry-men call too "sudden" a cross (for, without knowing the reason, they have found the evil of such often, and know it well), the *old white feather* may very probably reappear in all its original strength, though the new blood contained no tendency to it whatever. It is simply *the cross* of strange blood which gives the impulse to reversion. In the same way, to take the case mentioned just now, the pure white Spanish face being simply the result of assiduous breeding, and the most extreme care being needed for its preservation, the simple fact of crossing two entirely distinct strains gives the impulse to revert to the red face which belonged to the Minorca—in all probability the original breed from which it was derived." [1]

The value of cross-bred animals for breeding-purposes is diminished by this tendency to reversion, and the consequent loss of the power of transmitting any definite characters to their offspring.

It is generally admitted that, in the cases in which improvements are effected by crossing, the greatest change is produced by the first cross, and that the improvement resulting from a repetition of the process is uniformly slight.

This would undoubtedly be the case from the principles already presented : the greater the difference between the two parents, when one of them is prepotent in the transmission of its characters, the greater would be the resemblance of the offspring to the one, and the wider the divergence from the characters of the other parent; and, as the resemblance of the parents to each other would be gradually increased by successive crosses, the difference between the offspring and the inferior parent would as gradually diminish

It is claimed that the tendency to develop undesirable characters is increased by each successive cross ; [2] but the facts relating to this subject, in the history of the breeds that have been established by crossing, have not been recorded with sufficient exactness to furnish conclusive proof of the correctness of this opinion. It does not, however, appear to be improbable that such

---

[1] "Book of Poultry," p. 126.

[2] *Journal of the Royal Agricultural Society*, vol. xxiii., p. 352, vol. xx., p. 296 ; *Quarterly Journal of Agriculture*, vol. i., p. 178, vol. vii., p. 497; Sinclair's "Code of Agriculture," p. 95.

may be the case, from the fact that in each successive cross the relative potency of the pure-bred male would be diminished, as the females to which he is bred are improved in their characters.

On the other hand, it might seem probable that the improvement of the female would increase the tendency to a predominance of the desired characters, and thus intensify the influence of the male in the further improvement of the family. But cross-bred animals do not, as a rule, transmit to their offspring, as dominant characters, the peculiarities that they have derived from a superior breed, even when they appear to predominate in their organization.

When both parents are cross-bred animals—even in cases in which they both resemble the superior race from which they have derived their most obvious characteristics—the prevailing tendency in their powers of transmission is shown in the frequent recurrence of remote ancestral characters in their offspring.

In attempts to establish a new breed by crossing, this tendency to atavic transmission can only be overcome by a persistent and long-continued system of selection. "Changes, in fact, by crossing, are not to be effected in a short space of time; you must look forward to several years of constant exertion, before you can hope, in this manner, to alter your stock." [1]

Mr. Hogg says: "By the attempt which has been made to renovate the Scottish flocks by the Cheviot blood, we see the unexpected length of time necessary for completing and confirming the change.

"No class of animals which I am acquainted with

---

[1] Blacklock on "Sheep," p. 115.

adhere more tenaciously to family distinctions than sheep, and the longer the blood has been kept pure and unmixed with that of another family, the more powerfully do they resist a foreign connection; and in the case under our immediate consideration, the opposition to a coalition of natures is doubly powerful, as it is a forcing of the creature farther from a state of nature into one more artificial, more dependent, and more directly under the management of man. . . . After a course of twenty or twenty-five years, at which period the Cheviot peculiarities are got tolerably well-established, and every attribute of the old race seems to be completely suppressed, an individual lamb will, in some generations, still exhibit the wild air and shaggy coat of the ancient maternal line." [1]

Sir John Sinclair remarks that, " as to any attempt at improvement by crossing two distinct breeds or races, one of which possesses the properties which it is wished to obtain, or is free from the defects which it is desirable to remove, it requires a degree of judgment and perseverance to render such a plan successful as is very rarely to be met with." [2]

In summing up the arguments in favor of cross-breeding, Mr. Spooner says: " Although the benefits are most evident in the first cross, after which, from pairing the cross-bred animals, the defects of one breed or the other, or the incongruities of both, are perpetually breaking out, yet, unless the characteristics and conformation of the two breeds are altogether averse to each other, Nature opposes no barrier to their suc-

[1] *Quarterly Journal of Agriculture*, vol. i., pp. 176–179.
[2] " Code of Agriculture," p. 95.
10

cessful admixture; so that, in the course of time, by
the aid of selection and careful weeding, it is practi-
cable to establish a new breed altogether.[1] . . . Let us
conclude," he then says, "by repeating the advice
that, when equal advantages can be attained by keeping a pure breed of sheep, such pure breed should
unquestionably be preferred; and that, although crossing for the purposes of the butcher may be practised
with impunity, and even with advantage, yet no one
should do so for the purpose of establishing a new
breed, unless he has clear and well-defined views of
the object he seeks to accomplish, and has duly studied the principles on which it can be carried out, and
is determined to bestow for the space of half a life-
time his constant and unremitting attention to the
discovery and removal of defects."[2]

From the variety of improved breeds that can now
be obtained, adapted to almost every variety of climate
and system of management, it cannot be desirable to
attempt the formation of a new breed, as any special
qualities that may be required under particular circumstances can be more readily obtained by a modification of the characteristics of some existing breed
that approximates in its qualities to the proposed
standard.

As cross-breeding among cattle, sheep, and swine,
can only be recommended for the production of animals intended for the butcher, it may be well to consider some of the advantages arising from its judicious
practice for this purpose.

[1] *Journal of the Royal Agricultural Society,* vol. xx., p. 311.
[2] *Loc. cit.,* p. 313.

A large proportion of our farm-stock, for a long time to come, must necessarily consist of the so-called " natives " and the grades that have been produced from them by various crosses. These animals have the advantage of hardiness, but they are not good feeders, and do not arrive at maturity at as early an age as the modern pure breeds. When crossed with the best of the meat-producing breeds, they are at once improved in these important qualities in which they were before deficient, while in the quality of their flesh they may be equal, if not superior, to the more highly-bred animals of the pure breeds.

In the pure breeds in which the fattening qualities have been highly developed, an excessive activity of the formation of fat may be readily induced, in con-nection with a deficiency in lean flesh that diminishes the real value of the animal when it reaches its final destination on the block. The value of such animals consists in their ability to transmit to their offspring their general form, with the tendency to mature early and fatten rapidly.

When such animals are crossed upon natives or grades of inferior quality, it is not surprising to see in their offspring a quality of flesh that in its propor-tions of lean and fat is superior to that of either par-ent.

In speaking of a cross of the Lincoln and Leices-ter sheep, Mr. Mosscrop says : " The cross improves the size, the quantity of wool, and the quality of the mutton, although perhaps the distinguishing feature of the pure-bred Leicester—propensity to fatten at an early age—is somewhat impaired.

" The greater admixture of lean mutton, however, more than compensates for this, by giving a superior value to the carcass." [1]

In the cross of a superior breed upon the average stock of the farm, the best results can only be obtained by a better system of feeding than the original stock had been accustomed to. The old Scotch saying, that " the breed is in the mouth," expresses an important truth in stock management.

An increase in size and the ability to fatten rapidly would become a source of weakness rather than an advantage in animals that are unable to obtain a sufficient supply of food to give a full and active development of the system.

With every improvement in " blood" a corresponding improvement in feeding and management must be made, or Nature will surely thwart our plans by asserting her supremacy, and adapting the animal to the conditions in which it is placed.

It seems to be the prevailing opinion that the cross of a large male upon the females of a small breed is not advisable, on account of the difficulty in parturition which, it is presumed, would arise from the disproportionate size of the offspring.

This belief must be founded on theoretical considerations only, as difficulties of the kind do not often occur in actual practice. After an extended experience during the past ten years, in crossing rams of the Cotswold, Lincoln, and Southdown breeds on com-

---

[1] *Journal of the Royal Agricultural Society*, 1866, p. 329. For the advantages of crossing pure-bred and common swine, *see* Harris on " The Pig," p. 36.

mon-grade merino ewes, I have failed to meet with a single instance of difficult labor arising from such influence.

In many instances, for the sake of experiment, the smallest ewes were selected for crossing with the largest rams, but in no case was the labor unusually severe or protracted.

In establishing the "Charmoise" breed of sheep, M. Malingie-Nouel tell us that he frequently bred his mixed-blood ewes, that did not weigh more than twenty-five kilogrammes, to rams of the New Kent breed that weighed over one hundred kilogrammes, and that, in over two thousand cases of such contrast in the parents, he observed but a single "accident" from the disproportionate size of the lambs.[1]

The size of the young animal at the time of birth is evidently determined by the dam, while its development after birth may be influenced by the inherited qualities of either parent.

In the vicinity of large towns a peculiar system of cross-breeding is successfully practised in producing early lambs for the market.

Pure-bred rams of any of the improved English breeds may be used for this purpose, but the Southdowns, from their superior quality of flesh, are generally preferred.

As the rapid growth and development of the lambs is of the first importance, the ewes, which are selected from common-grade flocks, should be strong and healthy, although perhaps in low flesh, and, above all, good milkers.

[1] "Encyclopédie pratique de l'Agriculteur," vol. x., p. 595.

After the lambs are yeaned—about the first of March—the ewes should receive a liberal ration of grain and roots, to promote the secretion of milk, and, at the same time, improve their condition in flesh. At the age of two or three weeks the lambs will learn to eat meal and turnips, which should be placed in boxes to which they have ready access, in pens that their dams cannot enter.

With good shelter, which must of course be provided, and a system of high feeding, the lambs are ready for the market when from seven to ten weeks old, the highest prices being paid for the earliest lots.

If not too late in the season, from four to five dollars per head can be obtained for good ones, while inferior or late lots are sold at much lower prices.

After the lambs are sold, the ewes are fattened and sold for mutton, soon after shearing.

With the exception of the pure-bred ram, the entire flock is thus disposed of before the close of the year, the ewes frequently bringing one dollar per head more than their original cost.[1]

A new flock of ewes may then be procured, and the same method repeated.

The essentials of success in this method of management are, a high-bred ram that can impress upon his offspring the ability to mature early, high feeding to secure the greatest possible activity of this inherited tendency, and good shelter.

[1] For details of Mr. Taylor's system of management, see *The Cultivator*, 1862, pp. 77, 160, 174.

CHAPTER XI.

THERE are many theories in regard to the relative influence of parents upon their offspring, some of which, without sufficient reason, have been quite generally accepted as established physiological truths. The Highland Agricultural Society of Scotland, in 1825, awarded prizes to four essays[1] on this subject, that were presented for competition.

In the first volume of " Transactions " of the society, Mr. Boswell's essay is published in full, while abstracts only of the others are given.

Mr. Christian claims that " any hypothesis which would assign a superiority, or set limits to the influence of either sex in the product of generation, is unsound and inadmissible." His essay is but briefly noticed, yet, so far as the influence of sex alone is concerned, his position has not been successfully controverted. The theory advanced by Mr. Boswell,

[1] These essays were written by John Boswell, Rev. Henry Berry, Mr. Christian, and H. N. Dallas, in answer to the following question, which was presented by the society for discussion : " Whether the breed of live-stock connected with agriculture be susceptible of the greatest improvement from the qualities conspicuous in the male, or from those conspicuous in the female parent ? " (" Transactions of the Highland Agricultural Society," vol. i., p. 17).

that the male had the greatest influence in determining the characters of the offspring, became quite popular, from the apparent indorsement it received by the society, as it was the only one of the prize essays that was published in full.

As the cases cited by Mr. Boswell in proof of his theory are, almost without exception, susceptible of a different interpretation, the males used being more highly bred than the females, and therefore likely to be prepotent in the transmission of their qualities, his conclusions as to the superior influence of the male are not sustained by the evidence presented. Moreover, we find on record a large number of as striking instances of the resemblance of the offspring to the female parent, which, in themselves, must be fatal to the theory.

In the essay by Mr. Berry, a preponderance of the influence of either parent on account of sex is denied. The best-bred animal, however, is believed to have the greatest influence in determining the peculiarities of the offspring. Of the instances given by Mr. Berry to illustrate his position, the following are quoted on account of their bearing upon the theory of Mr. Boswell, already noticed : " The writer," says Mr. Berry, " has been for some years in possession of an improved breed of pigs, which are chiefly of a sandy or brown color. His sows of this breed crossed with common boars almost invariably produce litters of pigs of their own color. At the present moment he has a litter of eleven pigs from a brown sow of the improved breed, by a black-and-white boar of the common breed. The young pigs possess all the characteristics of the dam, and are of precisely the same color. In litters of pigs got by the improved boars from country sows, the color of the improved race also predominates in a similar manner. . . .

"The writer's brother was lately in possession of well-bred pigs, the most striking characteristic of which was a short, pricked ear. The produce of these with the large pendent-eared swine of North Wales was invariably similar in the ear to the higher-bred animal, whether male or female." [1] The number of cases in which the offspring resembles the male are undoubtedly more numerous than the cases of resemblance to the female, for the obvious reason that the males selected for breeding are, as a rule, more highly bred than the females with which they are coupled, and they have also more numerous offspring from which the cases of resemblance are selected.

Those who overlook this fact, as is evidently the case with Mr. Boswell, fall into the error of attributing the greater number of observed resemblances of offspring to the male parent to a predominating influence of sex.

The importance of securing males of the best quality, that from their superior breeding will be likely to be prepotent in the transmission of their characteristics, cannot be too strongly urged as one of the readiest means of improvement.

"It is generally admitted as a fact proved, that in the ox, horse, and other domestic animals, the purer or less mixed the breed is, there is the greater probability of its transmitting to the offspring the qualities

[1] "Transactions of the Highland Agricultural Society," vol. i., p. 41.

it possesses, whether these be good or bad. Economical purposes have made the male in general the most important, simply because he serves for a considerable number of females.

" The consequence of this has been that more attention has been paid to the blood or purity of race of the stallion, bull, ram, and boar, than to that of the females; and hence it may be the case that these males more frequently transmit their qualities to the offspring than do the inferior females with which they are often made to breed. But this circumstance can scarcely be adduced as a proof that the male, *cæteris paribus,* influences the offspring more than the female." [1]

Notwithstanding the predominant influence of the " best-bred " parent is the rule, the intensity of other conditions, in many cases, interferes and produces unexpected variations. If high breeding has been practised with reference to a single quality only, as, for instance, speed in the horse, and the qualities that give strength and constitution are neglected, the one-sided development of the animal may produce an unstable condition of the organization that is not favorable to uniformity in the transmission of the single character it is proposed to perpetuate.

Stonehenge, in noticing such exceptions to the general rule, says: " My own belief in this matter, founded upon observations made during a long series of years, on the horse as well as the dog, is, that no rule can be laid down with any certainty. Much

---

[1] Dr. Allen Thompson, article " Generation," in " Cyclopædia of Anatomy and Physiology," vol. ii., p. 472.

depends upon the comparative physical power and strength of constitution in each parent—even more, perhaps, than the composition of the blood.

" There have been many instances of two brothers being used in the stud, both among horses and greyhounds, in which one has almost invariably got his stock resembling himself, in all particulars, not even excluding color, while the descendants of the other have rarely been recognizable as his. Thus among horses the Touchstones have been mostly brown or dark bay, and as a lot have shown a high form as race-horses; while the Launcelots have been of all colors, and have been below mediocrity on the turf.[1] Several examples of the same nature may be quoted from among greyhounds, such as Ranter, Gipsey Prince, and Gipsey Royal, three brothers whose stock was as different as possible, but the fact is so generally recognized that it is not necessary to dwell upon it.

" Now, surely this difference in the power of transmitting the likeness of the sire, when the blood is exactly the same as it is observed to extend over large numbers, can only depend upon a variation in individual power. Not only does this apply to the males, but the females also show the same difference." [2]

After citing several other instances of such variations, Stonehenge concludes by recommending breeders to be guided by his thirteenth axiom, which is as follows : " The purer or less mixed the breed, the more likely it is to be transmitted unaltered to the

---

[1] Touchstone and Launcelot were full brothers.
[2] "The Horse," by Stonehenge, p. 147.

offspring. Hence, whichever parent is of the purest blood will be generally more represented in the offspring; but, as the male is usually more carefully selected, and of purer blood than the female, it generally follows that he exerts more influence than she does; the reverse being the case when she is of more unmixed blood than the sire." [1]

Stonehenge is undoubtedly correct in the opinion that this axiom, on the whole, is the safest guide to the breeder in making his selection of animals with reference to the relative potency that may be expected in the parents. The cases that he cites of full brothers transmitting different characters to their offspring, may be readily explained in accordance with principles that have already been presented, and it is, therefore, unnecessary to assume that they form exceptions to the general rule that the best-bred parent has the greatest influence upon the apparent characteristics of the offspring.

An examination of the pedigrees of the animals in question will show that several sub-families of the breed are represented in their ancestry, and we might reasonably expect that full brothers would inherit their leading characteristics from different branches of the family tree, which they in turn might transmit to their offspring.

A long course of breeding in the same definite direction, or within the limits of the same family, would be required to secure uniformity in the dominant characters transmitted by animals closely related.

---

[1] *Loc. cit.*, p. 139. *See* also *Journal of the Royal Agricultural Society*, vol. xxii., p. 9, and vol. i., p. 24.

M. Girou believed that the relative age and vigor of the parents determined their relative influence in moulding the characters of their offspring, and Stonehenge, in the paragraph above quoted, appears to be inclined to accept the theory as at least a plausible one. When all other conditions are equal, it may be true that the relative strength of constitution and physical vigor of the parents may, to some extent, determine their relative influence upon the dominant characters of their offspring, but there is no evidence that such influence is sufficiently intense to counteract or overcome, in all cases, the other causes of hereditary transmission. I have frequently observed instances of animals decidedly deficient in strength and vigor that were prepotent, even when coupled with those that were remarkable for their high constitutional development.

It cannot be doubted, however, that in cases of marked immaturity, or of an impaired condition of the system from extreme old age, the powers of transmission are less strongly marked than they are in the meridian of health and development; but these are extreme cases, that cannot be relied upon as indicating the normal laws of the function of reproduction.

The excessive use of the male impairs his powers of procreation, and undoubtedly diminishes the potency with which he transmits his qualities.[1]

When there is no marked prepotency on the part of either parent, the male offspring frequently resemble the father and the female resemble the mother.[2]

[1] "Massachusetts Agricultural Report," 1860, p. 172.
[2] Colin's "Physiologie comparée," vol. ii., p. 535 ; *Journal of the*

This is often the case in the transmission of disease, as has already been noticed in the chapter on hereditary diseases. Of two hundred and fourteen cases of consumption recorded by Lugol, one hundred and six were males and one hundred and eight females; of these, sixty-three males inherited the disease from their fathers and forty-three from their mothers: and sixty-one females inherited the disease from their mothers and forty-seven from their fathers.

Phillips gives the history of two hundred and sixty-four cases of insanity, from which we learn that, of one hundred and seventeen males, sixty-four inherited the disease from their fathers and fifty-three from their mothers; of one hundred and forty-seven females, eighty inherited the disease from their mothers and sixty-seven from their fathers.[1]

The peculiar horny excrescence on the skin of the porcupine-men (Lambert family), that was transmitted for several generations, was limited to the males of the family.[2]

The following case is reported by Dr. Stewart: " A single man aged twenty-four years, and the eldest son of a family consisting of two sons and two daughters, has well-marked *pityriasis versicolor* (a disease of the skin) affecting his chest, back, and arms, and which was first observed when he was about fourteen years of age; his brother, twenty years of age, now has

Royal Agricultural Society, vol. xvi., pp. 21–35; Ribot on "Heredity," p. 2; Darwin's "Animals and Plants under Domestication," vol. ii., p. 93.

[1] *Journal of the Royal Agricultural Society*, vol. xvi., pp. 21, 35.

[2] *British and Foreign Medico-Chirurgical Review*, April, 1861, p. 246.

it, though not to the same extent; and his father, paternal uncles, paternal grandfather, and seven male cousins on the paternal side, have all been similarly affected; the disease, strictly limited to the males, usually appeared in all of them at puberty, and disappeared about the age of forty or forty-five years; while the females of the family, although not suffering from it themselves, have transmitted it to their male children. Atavism through the opposite sex occurred when females intervened to check its direct transmission to males." [1]

This disease of the skin is not, however, confined to males, and cases are recorded in which it has been limited in a family to females. Mr. Sedgwick, in his remarks on color-blindness, says: "An analysis of upward of two hundred cases shows that the proportion of males affected is nine-tenths of the whole. But as I had occasion to state with reference to the same point in ichthyosis, this apparent preference for the male sex is not due to any peculiar inaptitude in the female sex to the defect; for when it has primarily affected the latter, its sexual limitation is complete, as in the interesting case published by Mr. Cunier, where the defect occurred in thirteen individuals belonging to five generations of one family, all of whom were females." [2]

Ribot remarks that "the resemblance between parents and children may undergo such metamorphoses as shall cause the child to resemble at one time the

[1] *British and Foreign Medico-Chirurgical Review*, April, 1863, p. 449.
[2] Ibid., April, 1861, p. 253; "Cyclopædia of Anatomy and Physiology," vol. iv., p. 1454.

father, and at another the mother. Girou de Buza-
reingues, in his work 'De la Génération,' containing
some curious facts observed by him, tells us that he
knew two brothers who in early life resembled their
mother, while the sister resembled the father.

"These resemblances were such as to strike all who
saw them. 'But now,' says he, 'and ever since their
youth, the two boys resemble the father, while the
daughter has ceased to be like him.'" [1]

Cases not unfrequently occur in which the disease
or defect is limited to one sex and transmitted by the
other, as in the case of ichthyosis above noticed.

"In the following cases of sebaceous tumors of the
scalp, which occurred in the practice of Dr. Henry
Stewart, and which were hereditarily limited to the
female sex, in the first case for ten and in the second
case for five generations, it will be observed that in
the first case limitation by age as well as by sex occurred, and also that some of the females derived the
inheritance from their paternal grandmother by atavic
descent, which affords an additional proof of the influence of sex, for, except when a male thus intervened to
arrest the appearance of the disease, the inheritance
was direct from parent to child. . . .

"The wife of a painter, aged fifty-four years, has
thirty-three sebaceous tumors of the scalp, none of
which are larger than a walnut; but thirteen years ago
nine sebaceous tumors, varying in size from a nutmeg
to a small orange, were excised by the late Mr. Morton, with considerable relief to the severe headaches
she had previously suffered from; her daughter and

[1] Ribot on "Heredity," p. 3.

her granddaughter are both affected by them, her sons are perfectly free; her brothers' daughters are troubled with them, as well as several female cousins of different degrees of relationship; her mother, grandmother and female relations backward for seven generations, were similarly affected; no female who had attained her tenth year of age was without them, while none of the males in the family had ever had them. . . .

"A single woman aged thirty years, the only child of her parents, and suffering from phthisis in the second stage, which she has inherited from her mother's family, has ten sebaceous tumors on the scalp, varying in size from a nutmeg to a pea, and which were first observed when she was about fifteen years of age; these tumors have been common to the females of her mother's family—her mother, maternal grandmother, maternal great-grandmother, and maternal great-great-grandmother, all had them, and so likewise have several female cousins on the mother's side of the first and second degrees of relationship; all the females, but none of the males in the family have suffered from them." [1]

Mr. Sedgwick also reports a case of warts on the hands of the mother during childhood (they disappeared after puberty), that were transmitted to her three daughters, while her two sons were exempt.

"In the report of hereditary malformation of the hands, affecting ten generations of the same family, it is stated that 'it was the women only who had the

---

[1] *British and Foreign Medico-Chirurgical Review*, April, 1863, pp. 450, 451.

misfortune of entailing the defect on their off-spring.'"[1]

A case of cleft iris (which is the analogue of hare-lip) is recorded by Mr. Sedgwick, who sums up the details as follows: "The chief points of interest in the case are—1. The transmission of the defect without its being shared in by the mother; 2. That, while two of her three sons had the defect, her three daughters were free from it; and, lastly, that the maternal grandfather, the maternal grand-uncle, the maternal uncle, and the son of this last named, all shared in the defect, which shows that the inheritance in this case extended to at least four generations."[2]

According to Mr. Wilde, "In a family of thirteen, in the county of Sligo, mute twins occurred twice, being the seventh and eighth births: in the former both children were mute females; in the latter, a male and female, the boy not mute. Of the entire thirteen births in that family, five were males, none of whom presented any defect; and eight were females, of whom seven were deaf and dumb; the order of the birth of the mutes being the third, fourth, fifth, eighth, ninth, and eleventh." The same author states that "the proportion of sexes of the deaf and dumb in England and Wales, where one in 1,738 of the inhabitants was affected, is 100 males to 82.9 females; in Scotland, where one in 1,340 of the inhab-

---

[1] *Edinburgh Medical and Surgical Journal*, vol. iv.. p. 252, 1808, as quoted in *British and Foreign Medico-Chirurgical Review*, July, 1861, p. 148, note.

[2] *British and Foreign Medico-Chirurgical Review*, April, 1861, p. 249.

itants was affected, is 100 males to 80.0 females; in Ireland, where one in 1,380 of the inhabitants was affected, is 100 males to 74.5 females; in Prussia, where one in 1,360 of the inhabitants was affected, is 100 males to 78.0 females—the last statement being taken from M. Baudin's statistics."[1]

Mr. Sedgwick reports a case of chronic rheumatic gout, which made its appearance "gradually, in a woman at the age of thirty years, and the joints of whose hands are now, at the age of forty-three years, much crippled and deformed; her mother, who died at the age of forty-six years, suffered greatly from chronic rheumatic gout, which had commenced thirteen years previously, and which had thoroughly crippled and deformed the joints of both hands; there were three brothers and four sisters, the eldest of whom was a brother aged forty-five years, all of whom have been free from any similar affection; and in another case which is at present under my observation, in which a girl, aged eighteen years, has the hands and feet dreadfully crippled by the same affection, which began at the age of fourteen years; her mother, maternal aunt, and maternal grandmother, have all suffered in the same way, while the males of the family have been exempt."[2]

Mr. Sedgwick reports the following case, which occurred in his own practice :

"Mrs. A——, under the age of forty years, and the mother of seven children, has not had for many

[1] From "Report of Census in Ireland," as quoted by Mr. Sedgwick in *British and Foreign Medico-Chirurgical Review*, July, 1861, p. 141.

[2] *British and Foreign Medico-Chirurgical Review*, July, 1862, p. 168.

years a sound tooth, the decay having begun very
early in life ; she has no brothers, but there are three
sisters, younger than herself, whose teeth are in a
similar state, and in all of whom the decay commenced
at a very early age; their mother was similarly affected in the teeth, and, like her four daughters, 'was a
martyr to the toothache.'

"Of Mrs. A——'s seven children, five are girls,
in four of whom, aged respectively sixteen, twelve,
nine, and seven years, the teeth began to decay at the
age of two years or soon afterward; in the youngest
girl, aged two years and a quarter, the teeth are not
decayed, but the dentition has been difficult.

"Of the two boys, the third and fifth children in
the order of birth, one died at the age of three years,
and the other has attained the age of four years, without any decay in their teeth. The father of these
children has sound teeth.

"Of Mrs. A——'s three sisters, the eldest has had
four children, two boys, aged fifteen and five years,
with sound teeth; and two girls, aged thirteen and
three years, with decayed teeth. The two other sisters of Mrs. A—— have no children." [1]

Mr. Sedgwick reports the following case of the
hereditary procreation of twins by one of his female
patients : " The mother, the maternal aunt, the maternal grandmother, and the maternal great-grandmother, have all had twins, but none of the sons in
these families have ever been known to transmit in
this way a double heritage, although some of them

[1] *British and Foreign Medico-Chirurgical Review*, April, 1863, p.
454.

have been twins, with twin-brothers, both of whom have, in some instances, married and had large families of children." [1]

Cases are on record of renal calculi inherited from the mother,[2] but as these might be attributed to intra-uterine development, the following case is of interest, as showing direct transmission of the disease by the father: "Mr. Squire gives the case of a still-born male child that he had the opportunity of examining, where the calices and the pelvis of the kidneys were filled with numerous uric-acid calculi, some of the size of small peas; the father had been operated on for stone, and was then passing uric-acid calculi by the urethra, and he was a continual sufferer from marked symptoms of the uric-acid diathesis."[3]

"Venette relates the case of two brothers who had an hereditary aversion to cheese; their mother had a decided taste for cheese, but the repugnance of the father was such that at only the smell of it he was ready to faint."[4]

The following case was observed by Michaelis: "Every one of the *male* posterity of a noble family at Hamburg, dating back to the great-grandfather, and remarkable for their military talents, was, at the age of forty years, attacked with madness; there remained only a single descendant, an officer, like his fathers, who was forbidden by the senate of the town to

---

[1] *British and Foreign Medico-Chirurgical Review*, July, 1861, p. 148.

[2] "Cyclopædia of Anatomy and Physiology," vol. ii., p. 336.

[3] *British and Foreign Medico-Chirurgical Review*, July, 1863, p. 166.   [4] Ibid., p. 164.

marry: the critical age arrived, and he lost his reason." [1]

The numerous cases in which peculiarities belonging exclusively to one sex are transmitted by the other are of particular interest, as they illustrate the manner in which resemblances are sometimes transmitted by atavic descent.

Mr. Talcott reports the following case. He says: "I had a fine cow with nice bag and teats, which I took to a bull in the neighborhood, and the produce was a heifer-calf, which was raised because of the good milking-qualities of her dam; but when she became a cow, instead of any of the good qualities of her dam as was expected, her bag and teats were more in contrast (*sic*) with that of a sheep than of a good dairy-cow. I then began to investigate the cause, and found that the heifer was the counterpart of the dam of the bull, she being an ordinary cow with a small bag and still smaller teats, and from that time to this I have found that too frequently that is the case, especially if the bull was from such a stock or family of light milkers that it was not desirable to perpetuate them. I remember distinctly the first pure-bred Short-Horn bull I ever had, that the bag of his dam was the largest in the hind-quarters, consequently that she gave most milk from the hind-teats, and that quality was transmitted to the majority of his heifers when they came to be cows, their bags tending largely in the hind-quarters. And I think, from such observations, that

[1] Prosper Lucas, "De l'Hérédité naturelle," vol. i., p. 255; as quoted by Sedgwick, in *British and Foreign Medico-Chirurgical Review*, April, 1863, p. 473.

there can be no doubt that such is the case generally."[1]

It is well known to the breeders of Ayrshire cattle that the sire has an important influence upon the form and functional activity of the udder; and the position and development of the false teats of the bull are believed to furnish an indication of the milking qualities he will be likely to transmit.

In the large number of grade Ayrshires that I have bred for dairy-purposes, the udder, in most instances, has resembled the family type of the sire in form and general proportions. The males of the dairy-breeds, generally, are prepotent in the transmission of the characteristics of the females of their race.

"It is well known, for example," says Mr. Sedgwick, "that the supply of milk by cows is hereditarily influenced by the bulls rather than by the cows from which they are directly descended, and that the character of the secretion, as regards both the quantity and the quality of the milk, is chiefly derived from the paternal grandmother, by atavic descent" (Burdach, "Traité de Physiologie," vol. iii., page 117; and Girou, *op. cit.*, page 127); "and as we descend still lower in the scale, we find, for example, in the case of insects, evidence more or less decisive in favor of the transmission by either sex of the distinctive peculiarities of the other; while the capability of both sexes in the human race to transmit disease by atavic descent is occasionally illustrated by the occurrence of cases in which the transmission is effected by a male

[1] *Country Gentleman*, April, 1865, p. 236. For other similar cases, see *Country Gentleman*, 1873, p. 42.

and a female branch of the same family, as in the following case, related by F. Meckel" (*Lancet*, 1829 -'30, vol. i., page 792), "in which the modified influence of sex is associated with atavism of unequal remoteness; 'a man whose palate was entire, but uneven, as if cicatrized, had, by a perfectly healthy wife, seven children, of whom the four boys were well formed, but the three girls had hare-lip and divided palate. His mother's sister had also seven children, five sons and two daughters, of whom the former were all similarly deformed.' " [1]

In the two following cases, the one of a disease and the other of a congenital defect of the male organs of generation, the female parent transmits to her offspring peculiarities that she could not herself be affected with.

Sir Henry Holland reports a case "of hydrocele occurring in three out of four generations in one family, the omission adding to the singularity of the fact from its depending on a female being third in the series, in whose son the complaint reappeared." [2]

Many cases of hereditary hypospadias (a defect of the male urethra) are on record.

"In a case observed by Meckel, it appears that a woman, born of a family which presented many examples of hypospadias, gave birth to two boys affected with the deformity." [3]

---

[1] *British and Foreign Medico-Chirurgical Review*, July, 1863, p. 193.

[2] *Edinburgh Medical Journal*, 1858-'59, p. 501; quoted in *British and Foreign Medico-Chirurgical Review*, July, 1861, p. 148.

[3] *British and Foreign Medico-Chirurgical Review*, July, 1863, p. 173.

A case is reported "of the total absence of the uterus in three out of five daughters in the same family." This is supposed to be an instance of collateral inheritance through the males of the family, as it could not, of course, be directly transmitted.[1]

Mr. Sedgwick, on the authority of Dr. Russell, of Birmingham, gives the following case of hereditary obesity, limited to the male sex: "The first is the case of a very stout and flabby man, with copious deposit of fat, and symptoms of fatty heart; he has four brothers and one sister: the sister is thin, while one of his brothers is as large as himself, and the three others are larger; his father, paternal uncle, and paternal grandfather were large and fat men; his mother was of medium size, and his maternal grandmother was tall and thin. The second case is that of a very stout man, aged twenty years, with a very large amount of subcutaneous fat, and symptoms of a fatty heart; he has had ten brothers and sisters, of whom only two brothers and two sisters are living; the two brothers are even fatter and heavier than he is, while the two sisters are of only medium size; his father was, as a young man, always very fat, and other male relations in the family are also large-made and fat."[2]

"A sporting-dog, the issue of a setter mother and a spaniel father, was coupled with a setter bitch, and the male offspring were spaniels, like the paternal grandfather, and resembled him in their hair, while

---

[1] *British Medical Journal*, October 5, 1861, p. 359; as quoted by Sedgwick, *loc. cit.*, p. 171.

[2] *British and Foreign Medico-Chirurgical Review*, July, 1863, p. 168.

11

the female offspring were setters, having the color of their mother.[1]

"A family of Angora cats, of which the mother is white and deaf; the father, which hears, is white and black; all the kittens which are born white are deaf as the mother, those which resemble the father are not so."[2]

In the case of Augustin Duforet, already referred to, the malformed digits in the third generation were inherited by the twelve sons, while the seven daughters were exempt, and the same sexual limitation occurred in the second generation with a single exception.[3]

Mr. Dallas,[4] in one of the Highland Agricultural Society essays, already mentioned, advances the theory that the male has the greatest influence on the external appearance of the offspring, and the female on the internal qualities; and this division of influence he accounts for on the supposition that the seminal fluid of the male invests the ovum, and thus forms its outer envelope, while the germ itself, from which the internal structures are formed, is furnished by the

---

[1] Sedgwick, loc. cit., April, 1863, p. 451, who quotes from "De la Génération," by Girou, p. 123.

[2] Sedgwick, loc. cit., p. 458; on the authority of M. Bouyer-Desmortiers.

[3] Quite a number of instances of sexual limitation of hereditary characters may be found among the cases cited to illustrate other forms of heredity in the preceding chapters. A summary of the facts presented by Dr. Prosper Lucas and Mr. Sedgwick, with additional cases, has been given by Darwin, in "Animals and Plants under Domestication," vol. ii., pp. 93, 94.

[4] "Transactions of the Highland Society," vol. i., p. 43.

female. As this physiological exposition, on which the theory is based, is a pure assumption, in direct conflict with the known facts of embryology, the essay may be passed without further comment.

A modified form of this theory has been elaborated by Mr. Walker,[1] who draws most of his illustrations from the human family; and, more recently, Mr. Orton[2] has advanced the same theory, in its applications to stock-breeding.

As a large proportion of modern writers on the physiology of breeding have quoted the arguments of Walker and Orton with approval, their theories have assumed an importance that is not warranted by their real merits.

Mr. Walker enunciates his first law as follows: " Where both parents are of the same variety, . . . *one parent communicates the anterior part of the head* (and I believe the upper middle part also), *the osseous or bony part of the face, the forms of the organs of sense* (the external ear, under lip, lower part of the nose, and eyebrows, being often modified), *and the whole of the internal nutritive system* (the contents of the trunk, or the thoracic and abdominal viscera, and consequently the form of the trunk itself, in so far as that depends upon its contents). The resemblance to that parent is, consequently, found in the forehead and the bony parts of the face, as the orbits, cheekbones, jaws, chin, and teeth, as well as the shape of the organs of sense, and the tone of the voice. . . . *The other parent communicates the posterior part of*

[1] " On Intermarriage," 1839.
[2] " On the Physiology of Breeding," two lectures, 1855.

the head (and I believe the lower middle part also), *the cerebral, situated within the skull, immediately above its junction with the back of the neck, and the whole of the locomotive system* (the bones, ligaments, and muscles, or fleshy parts). The resemblance to that parent is, consequently, found in the back-head, the few more movable parts of the face, as the external ear, under lip, lower part of the nose, eyebrows, and the external forms of the body, in so far as they depend on the muscles, as well as the form of the limbs, even to the fingers, toes, and nails." [1]

" It is a fact," says Mr. Walker, " established by my observations, that, in animals of the same variety, *either the male or the female parent* may give *either series of organs,* as above arranged—that is, *either* forehead and organs of sense, together with the vital and nutritive organs, *or*, back-head, together with the locomotive organs." [2]

" The second law, namely, that of crossing, operates where *each parent* is of a *different breed,* and when, supposing both to be of equal age and vigor, the *male* gives the back-head and locomotive organs, and the *female* the face and nutritive organs." [3]

" The third law, namely, that of in-and-in breeding, operates where *both parents* are not only of the same variety, but of the *same family in its narrowest sense,* and when the *female* gives always the *back-head and locomotive organs,* and the *male the face and nutritive organs*—precisely the reverse of what takes place in crossing." [4]

[1] Walker on " Intermarriage," pp. 142, 143.
[2] *Loc. cit.,* p. 146.　[3] Walker, *loc. cit.,* p. 184.　[4] Ibid., p. 204.

According to Mr. Orton,[1] " the male animal influences especially the external, and the female the internal, organization of the offspring. The outward form, general appearance, and organs of locomotion, are chiefly determined by the male; the vital organs, size, general vigor, and endurance, by the female."[2]

As stated by Goodale, he maintains that " the male parent chiefly determines the external characters, the general appearance, in fact, the outward structure and locomotive powers of the offspring, as the framework, or bones and muscles, more particularly those of the limbs, the organs of sense, and skin; while the female parent chiefly determines the internal structures and the general quality, mainly furnishing the vital organs, i. e., the heart, lungs, glands, and digestive organs, and giving tone and character to the vital functions of secretion, nutrition, and growth."[3]

Mr. Spooner says: " The most probable supposition is, that propagation is done by halves, each parent giving to the offspring the shape of one-half of the body. Thus the back, loins, hind-quarters, general shape, skin, and size, follow one parent; and the fore-quarters, head, vital and nervous system, the other; and we may go so far as to add that the former, in the great majority of cases, go with the male parent, and the latter with the female."[4]

[1] Not being able to refer directly to Mr. Orton's original paper, the statements of his opinions are quoted from the *Journal of the Royal Agricultural Society*, vol. xvi., p. 43; Goodale's " Principles of Breeding," pp. 73–79; and *Journal of the Highland Agricultural Society*, 1857–'59, pp. 19–22. [2] *Jour. of the Royal Agricul. Soc.*, vol. xvi., p. 43.
[3] " Principles of Breeding," p. 75.
[4] *Journal of the Royal Agricultural Society*, vol. xx., p. 295.

Mr. Spooner adds,[1] however, that "the size is governed more by the male parent;" while Mr. Orton is equally positive that the size must follow the female parent.[2]

It will be noticed that the advocates of the half-and-half theory of generation do not agree in many particulars as to the supposed division of parental influence, and this in itself may fairly be urged as an objection to the theory.

When the offspring in external form resembles one parent, it does not follow that the internal or vital organization is derived from the other parent, and the advocates of this theory have failed to produce any evidence that can possibly warrant such a conclusion.

In crossing a pure-bred male, of any of the improved meat-producing breeds, upon native or cross-bred females, the sire is not only prepotent in determining the external form and characters of the offspring, but he has also a predominant influence upon the organs of nutrition, as is shown in the uniform superiority of the grade animal to its dam, in size, feeding quality, and early maturity.

Instead of a limitation of the influence of each parent to a particular set of organs, we find the parent that is prepotent in the transmission of its characters has a controlling influence upon the internal as well as the external organization of the offspring.

Physiological objections may likewise be made to the classification and presumed origin of the various organs of the body, in each of the three forms of the theory under consideration.

[1] *Loc. cit.*, p. 295.   [2] "Principles of Breeding," p. 78.

In an early stage of the development of the germ, a blastodermic membrane is formed, from which the embryo is developed. This blastodermic membrane is soon separated into two layers, which are designated as the external and internal layers of the germinal or blastodermic membrane. "According to the most recent observations, the main portion of the external layer, sometimes called the serous layer, simply forms a temporary investment for the rest of the vitellus (yolk), and is not developed into any part of the embryon. The internal layer, called the mucous layer, is developed into nothing but the epithelial lining of the alimentary canal. There is a thickening of both of these layers at the line of development of the cerebro-spinal system, with a furrow, which is finally inclosed by an elevation of the ridges and their union posteriorly, forming the canal for the spinal cord. As the spinal canal is thus developed, a new layer is formed by a genesis of cells from the internal surface of the original layer and the opposite surface of the internal or mucous layer. This layer of new cells may be termed the intermediate layer, and it is from this that nearly all the parts of the embryon are developed.

"To summarize the development of the layers just mentioned, we may state that the external layer is a temporary structure; the internal layer is very thin, and is for the development of the epithelial lining of the alimentary canal; and the most important structure is a thick layer of cells developed from the opposite surfaces of the external and the internal layers, and situated between them, called the intermediate layer; and it is from these cells that the greatest part of the embryon is formed." [1]

As there is a tendency to a subdivision of the three layers mentioned, some modern physiologists include in the external layer the upper surface of the intermediate layer, while the lower surface of the intermediate layer is included in the inner layer of the blastodermic membrane.

This does not, however, involve any difference of opinion as to the parts of the germinal membrane that are developed into the different organs of the body. Dr. Marshall, who adopts the latter classification, says : " From the upper *external* or serous layer, also named the *sensorial* layer, are developed, along its axial portion, the cerebro-spinal nervous axis and the organs of the senses, and, from its lateral portions, the cuticle or outer skin, with its epidermic appendages, the feathers, bill, and claws, and, in the mammalia, the nails and hairs ; lastly, the sebaceous and sudoriferous cutaneous glands, and the Meibomian, ceruminous, and mammary glands.

" From the *middle* layer, also called the *motoriosexual* layer, are developed, by complicated metamorphoses of its substance, the bones, the muscular system, the peripheral spinal nerves, the sympathetic nerves, the heart, blood-vessels, and lymphatic system, the so-called ductless glands, and the reproductive organs ; also, next to the external layer, the true skin, and, next to the internal layer, the muscular and submucous coats of the alimentary canal.

" Lastly, from the internal layer, also called the

[1] Flint's "Physiology," 1875, vol. v., p. 360.

mucous or *intestinal* layer, are developed the epithelial lining of the alimentary canal, and all its glandular extensions, such as the mucous, the gastric, and intestinal glands, the pancreas, and the liver, also the lungs and respiratory passages, and the urinary apparatus, including the bladder, ureters, and kidneys." [1]

From this outline of the origin of the different organs in the development of the embryo, it will be seen that the classification of organs made by Messrs. Walker, Orton, and Spooner, is not in accordance with their true relations in the process of embryological development.

As both locomotive and nutritive organs are developed from the middle germinal layer, which is derived from a cell outgrowth of the external and internal layers, it does not seem probable that either group of organs is produced by the exclusive influence of one parent.

Many of the arguments advanced in favor of this theory are drawn from fancied analogies that are not in harmony with well-established facts.

" It is clear," says Mr. Walker, " that the whole nutritive system, chiefly contained within the trunk, is naturally connected with the senses of taste and smell, which are the guides to the supply of its wants as to food and drink, and therefore the senses contained in the face (and consequently the observing faculties dependent on these senses and contained in the forehead) *ought* to accompany the nutritive system."

And, by a similar process of reasoning, he concludes that " the back-head, containing both the organ

[1] " Outlines of Physiology," p. 954.

of will and the posterior masses of the brain—the seats of desire or aversion by which will is excited— *ought* to accompany the locomotive system, not merely in the greater masses of the figure, but even in the muscles of the face." [1]

It may be that this *ought* to be the case, but Nature has, unfortunately for herself or the theory, developed the structures so closely associated from quite different portions of the blastodermic membrane of the embryo.

As the peculiarities of hybrids are relied upon as furnishing the most conclusive evidence of the truth of this theory, an examination of this part of the argument will be of particular interest.

" The mule," says Mr. Orton, " the produce of the male ass and the mare, is essentially a *modified ass;* the ears are those of an ass somewhat shortened; the mane is that of the ass, erect; the tail is that of an ass; the skin and color are those of an ass somewhat modified; the legs are slender, and the hoofs high, narrow, and contracted, like those of an ass; in fact, in all these respects it is an ass somewhat modified. The body and barrel of the mule are round and full, in which it differs from the ass and resembles the mare. The hinny [2] (or muto), on the other hand, the produce of the stallion and she-ass, is essentially a modified horse; the ears are those of a horse somewhat lengthened; the mane flowing; the tail is bushy, like that of a horse; the skin is finer, like that of a horse; and the color varies also like the horse; the

---

[1] Walker on " Intermarriage," pp. 143, 144.
[2] *Bardeau* of the French.

legs are stronger, and the hoofs broad and expanded, like those of a horse. In fact, in all these respects it is a horse somewhat modified.

"The body and barrel, however, of the hinny are flat and narrow, in which it differs from the horse and resembles its mother, the ass. It is clearly evident that these two hybrid animals have followed the male parent in all his external characteristics.

"In two respects there is, however, a striking departure from him. First, in size they both follow the female parent, the mule being in all respects a larger and finer animal than its sire, the ass; while the hinny is just the reverse, being flat and narrow. In this respect the mule is just the reverse of its sire, the ass, while the hinny is just the reverse of its sire, the horse; while both, also, in this respect (the body and barrel) resemble their female parent."[1]

Mr. Orton adds: "The mule *brays*, while the hinny *neighs*. The why and wherefore of this is a perfect mystery, until we come to apply the knowledge afforded us by the law I have given. The male gives the locomotive organs, and the muscles are among these; the muscles are the organs which modulate the voice of the animal; the mule has the muscular structures of its sire, the ass; the hinny has the muscular structures of its sire, the horse; the organs of voice in the former are those of its sire, the ass, hence it brays; the organs of voice of the latter are those of its sire, the horse, hence it neighs."[2]

[1] Quoted from the *Journal of the Royal Agricultural Society*, 1857–'59, p. 21.

[2] *Journal of the Highland Agricultural Society*, 1857–'59, p. 22.

There is a substratum of truth in these statements that, at the first glance, gives a plausibility to the argument in favor of this theory—which, however, disappears when all the facts are presented in their true relations.

Without noticing the fallacies in the statement in regard to the resemblance of the mule to its "sire, the ass," it may be admitted that the ass is prepotent in the transmission of its characters when bred with the mare, and that the mule consequently presents a stronger resemblance to its sire than to its dam.

According to M. Colin, who is one of the highest authorities on the comparative anatomy and physiology of domestic animals, the hinny resembles the ass more closely than it does the horse. It has a finer head than the mule, and in the mane and tail more nearly resembles the horse; but, in general form and size, in peculiarities of the nostril, the withers, the back, the legs and feet, and other minor peculiarities, it presents a stronger resemblance to the dam than to the sire. "On the whole," says M. Colin, "in the produce of the two species, the ass and the horse, it is undoubtedly the influence of the ass that predominates in the transmission of the external form, the constitution, and the disposition."[1]

After noticing the results of the cross of the zebra and the horse, the ass and the zebra, and the hemione and the ass, M. Colin concludes as follows: "In examining the mules of solipeds, we see that if the mule, properly so called, resembles the ass, its father,

[1] "Physiologic comparée des Animaux domestiques," tome ii., p. 537.

in general form, in the head, the mane, the back, the tail, the legs, and the feet; the hinny (*bardeau*) that, on the whole, differs but little from the mule, resembles the ass, its mother, in the large number of its points, the head and the mane excepted. If, then, it is true that the first (the mule) derives its form from its father, it is equally true that the second (the hinny) derives its form from the mother, and, if the parents transmit their form to the anterior part of the body, they do the same for the posterior part. Consequently what the father gives to the mule, the mother, with but slight variation, gives to the hinny.

"In the second place, if the mule derives from the ass, its father, its constitution, strength, hardiness, and disposition, the hinny derives the same characters from the ass, its mother, as there is a stronger resemblance of the two hybrids in these characters than there is even in external conformation; and, finally, if the mule derives its size from the mother, why is it not her equal in this respect? and, if the hinny derives its size from the mother, why does it exceed her in size?"[1]

Some of the advocates of the theory under review admit that, so far as size of the offspring is concerned, a preponderating influence cannot be exclusively attributed to either parent. Mr. Spooner says: "How often do we find that, in the by no means infrequent case of the union of a tall man with a short woman, the result in some instances is that all the children are tall, and in others all short; or, sometimes, that some of the family are short and others tall! Within our

---

[1] Colin, *loc. cit.*, p. 539.

own knowledge, in one case where the father was tall
and the mother short, the children, six in number, are
all tall.   In another instance, the father being short
and the mother tall, the children, seven in number,
are all of lofty stature.   In the third instance, the
mother being tall and the father short, the greater
portion of the family are short."[1]

The resemblance of the hinny to the mule, noticed
by M. Colin, has likewise been observed in this country.   Mr. B. F. Cockrell, of Nashville, Tennessee,
says: "In the year 1850 I bred a dozen jennets to a
thorough-bred son of imported Priam.   The following
spring I had six mules foaled.   In 1851 I again bred
these same jennets to the same stallion, and had four
mules foaled, three of which lived and attained maturity.   I often asked visitors to point out the hinnies
from the mules (there being other mules on the plan-
tation), and in no instance did I ever find a man that
could distinguish them from other mules.

"I shipped them, with forty other mules, to my
father's cotton-plantation in Mississippi, where they
did the same routine of duty with the other mules,
and remained in all respects perfectly *incognito* as to
color, feet, head, voice, and size, to their death."[2]

A correspondent of *The Country Gentleman*, in
reference to the hinnies bred by Mr. Cockrell, says:
"I have seen but one hinny, to my knowledge.   It
was before a wagon, alongside of a mule, last year.   I
examined it closely, but was unable to see any structural peculiarities to distinguish one from the other.   It

[1] *Journal of the Royal Agricultural Society*, vol. **xx.**, p. 295.
[2] *The Country Gentleman*, 1876, p. 170.

was colored more like the jennet than most mules, but I have seen many with similar markings, and many more like the horse in style and finish than it was."[1]

The difference in the voice of the mule and the hinny, that Mr. Orton urges as an important part of his argument, requires a passing notice. The larynx is the organ of voice in mammals, and modifications of its form give rise to the various sounds emitted by different species of animals. In the ass the vocal ligaments are inserted in an arched cavity. " On each side of this cavity are two circular apertures, which lead to two large sacs situated behind the mucous membrane, between the vocal ligaments and internal surface of the thyroid."[2]

The characteristic bray of the ass is produced by this peculiar conformation of the larynx; and the mule, inheriting the same structure, is endowed with a similar voice. As the larynx is not developed from the layer of the blastodermic membrane that gives rise to the locomotive organs, the argument of Mr. Orton, in regard to the peculiarities of the voice of the mule and the hinny, is without foundation.

"A cross between a male wolf and a bitch," says Mr. Orton, " illustrates the same law, the offspring having a markedly wolfish aspect, skin, color, ears, and tail. On the other hand, a cross between the dog and female wolf afforded animals much more dog-like in aspect, slouched ears, and even pied in color.

---

[1] *Loc. cit.*, p. 170.
[2] "Cyclopædia of Anatomy and Physiology," vol. iv., p. 1492; *Journal of the Highland Agricultural Society*, 1857-'59, p. 22.

"If you look to the descriptions and illustrations of these two hybrids, you will perceive at a glance that the doubt arises to the mind in the case of the first, 'What genus of *wolf* is this?' whereas, in the case of the second, 'What a curious *mongrel dog!*'"[1]

If this statement of the relative influence of the parents, in the case of a cross between the wolf and the dog, could be made to agree with known facts, it would furnish a very strong argument in favor of the half-and-half theory.

Buffon, however, mentions "the very conclusive case of a she-wolf which had two cubs, a male and a female, to a setter-dog. The male resembled the father in external appearance, except that the ears were pointed, and the tail like that of the wolf; the female resembled the mother, and had all her characteristics, with the exception of the tail, which was that of the dog." The same author informs us that "the produce of a dog and a she-wolf sometimes bark and sometimes howl; and the produce of a bitch-fox and a dog, according to Burdach, barked like a dog, though somewhat hoarsely, and howled like a wolf when it was hurt. A similar remark has been made by all who have attended to cross-breeding in birds; the hybrid of the goldfinch and the canary has the song of the goldfinch, mingled with occasional notes of the canary, which seem perpetually about to gain the predominance."[2]

On the whole, it must be admitted that the evi-

---

[1] *Journal of the Royal Agricultural Society*, vol. xvi., p. 44; Goodale's "Principles of Breeding," p. 78.

[2] *Journal of the Highland Agricultural Society*, 1857-'59, p. 22.

dence relating to hybrids tends to disprove the theory under consideration.

If the male furnishes the external characters, the color of the offspring should, as a rule, follow that of the sire. Mules are, however, varied in color; the white, the gray, the iron-gray, the black, the dun-colored, the spotted, and the cream-colored, are of common occurrence.[1]

"A cow of the Swiss race, having a white skin, spotted with red, is mentioned by M. Girou as having produced five calves, only one of which, a female, resembled the bull, and four males which were like their mother, both in the ground-color of the skin and the distribution of the spots.

"Instances of this nature have been observed by every one in possession of a herd of cattle; it is never expected that the produce should always resemble the bull in color; even though his color may predominate in a herd, sufficient variety never fails to appear. Black-and-white kittens are every day produced from cats one of which is wholly black and the other wholly white. A black buck and a white doe have produced at one time a black-and-white fawn, and, at another time, one entirely black, except a spot above the hoof."[2]

The white feet and face of the celebrated horse Dexter are characteristic of his dam and grandam, who transmitted the same marks, with great uniformity, to their offspring.[3]

---

[1] *See* "The Mule," by Riley, pp. 22, 40.

[2] *Journal of the Highland Agricultural Society*, 1857–'59, p. 23.

[3] *National Live-Stock Journal*, 1876, p. 57.

Mr. Roberts reports the case of a horse with two curbs. The sire was free from defects; but a sister in the same stable had two curbs; their dam had two curbs, and a foal of hers by another horse had also two curbs, showing conclusively that the defect was transmitted by the dam.[1]

The statements of Mr. Walker have been so often quoted, and his theory so generally accepted, that we must be permitted to quote some of his cases and his inferences from them : "Of the power of the horse to communicate, in a cross, his skeleton, and therefore his locomotive system generally, or, in other words, his general shape and character, Mr. Knight gives an interesting example."[2]

Then follows Mr. Knight's case : "I have obtained offspring," he says, "from Norwegian pony mares and the London dray-horse, of which the *legs are preternaturally short*, and the shoulders and body preternaturally deep, and the animal of course preternaturally strong. . . . The offspring of my Norwegian mares, *as always happens* in similar cases, *had legs as short as their mothers* at birth ; but the male parent, the dray-horse, caused their legs to grow greatly stronger, and their joints and bodies generally much larger, *although* the legs remained short."

"Thus in equine crosses," says Mr. Walker, in the paragraph immediately following the quotation from Mr. Knight, "the male gives the locomotive system, the female the vital one."

A theory founded on such inferences, from such

---

[1] "The Horse," by Youatt, p. 35.

[2] Walker on "Intermarriage," p. 187.

facts, would hardly seem to require further notice, but we must not overlook some of its applications that are of practical interest. "A and B," says Mr. Walker, "who are more or less perfectly crossed, may have very different vital and locomotive systems: of their immediate progeny, C may have the vital system of A and the locomotive system of B; and D may, on the contrary, have the locomotive system of A and the vital system of B (for, in a feeble or imperfect cross, such variation may occur); and, of the progeny of these last, E may have from C the vital system of A, and from D the locomotive system of A; and F may have from C the locomotive system of B, and from D the vital system of B. Thus A and B may be reformed in the third generation." [1]

The statement, in its simplest form, is that the grandchildren E and F are identical with the grandparents A and B in organization, while the parents C and D are each one-half A and one-half B.

The absurdity of this proposition will be readily seen in its application to a particular case.

According to the theory a pure Devon bull, bred to a pure Short-Horn cow, may produce a bull-calf with the external organization of the Devon and the internal organization of the Short-Horn; the same pair might also produce a heifer-calf with the external organization of the Short-Horn and the internal organization of the Devon. With this division of the organization it follows that, if the two cross-bred animals are bred together, the offspring in one instance

---

[1] Walker, *loc. cit.*, p. 199. A similar statement in regard to the cross of the Arab horse may be found in the same work, p. 183.

may inherit the external Short-Horn characters of one
parent and the internal Short-Horn characters of the
other parent, and thus be a pure Short-Horn ; and, in
another instance, the external characters of the Devon
may be inherited from one parent and the internal
characters of the Devon from the other parent, and
thus be a pure Devon. In other words, a pure Devon
and a pure Short-Horn may, according to the theory,
be produced from the same pair of cross-bred animals,
which is not only absurd, but in direct conflict with
all the known facts in cross-breeding. If the external
characteristics are represented by the numerator of a
fraction and the internal characteristics are represented by the denominator, the inherited characters of the
offspring, according to Mr. Walker, may be clearly
represented in the following diagram, in which D
stands for Devon and S for Short-Horn.

Original Parents.   First Produce.   Second Produce.

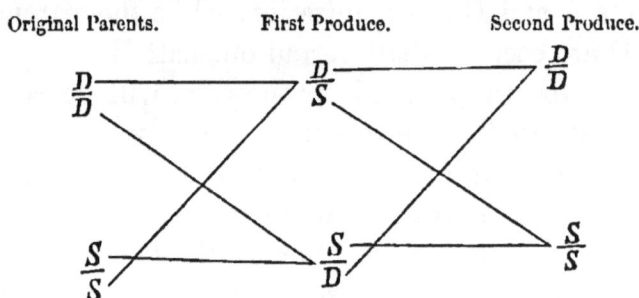

$$\frac{D}{D} \qquad \frac{D}{S} \qquad \frac{D}{D}$$

$$\frac{S}{S} \qquad \frac{S}{D} \qquad \frac{S}{S}$$

Mr. Orton, who was apparently aware of the in-
consistency involved in the application of this theory
to extreme cases, makes an admission that is fatal to
the theory he advocates. He says : " I do not mean
to imply or state that in all cases the law operates
with the precision I have above stated, for there are
certain controlling influences which *confuse* and *mod-*

ify, and in some cases *almost seem to set aside the law*. I do not mean it to be inferred that either parent gives either set of organs uninfluenced by the other parent, but merely that the leading characteristics and qualities of both sets of organs are due to the male on the one side and the female on the other, the opposite parent modifying them only."[1]

It must be obvious, from the facts already presented, that the half-and-half theory of generation cannot be true.[2]

The characteristics of one parent may sometimes be transposed, in some unaccountable manner, through the supplementary influence of the other parent, as in the following remarkable case reported to me by Dr. H. B. Shank, of Lansing, Michigan: A white cat with a small black patch, consisting of a few hairs, on her forehead, had kittens by a tomcat that was entirely black. The kittens were all black, with the exception of a small patch on the forehead, which was white. The white patch on the kittens occupied the same position, and it was also of the same size, as the black patch on the forehead of the mother.

The relative influence of parents upon their offspring evidently depends upon conditions that cannot in all cases be determined. When the characteristics of one parent have been fixed by the inheritance of the same peculiarities for many generations, it will undoubtedly prove to be prepotent in the transmission

---

[1] *Journal of the Highland Agricultural Society*, 1857–'59, p. 25.

[2] Darwin's "Animals and Plants under Domestication," vol. ii., pp. 88, 432; "Heredity," by Ribot, p. 166; *Journal of the Highland Agricultural Society*, 1857–'59, p. 21.

of its characters if the other parent has a less stable
organization, but this will not prevent the inheritance
of the peculiarities of both parents that are not included in the dominant characteristics.

The cases of cross-heredity, or the transmission by
one sex of the peculiarities of the other, in connection
with cases like the last above cited, are, however, sufficient to show that there are laws governing the transmission of characters that, in the present state of
knowledge, we are unable to define.

CHAPTER XII.

THE influence of the male in the process of procreation is not limited to his immediate offspring, but extends also, through the female that he has impregnated, to her offspring by another male.

Paradoxical as this statement may appear, there are many well-authenticated cases on record that cannot be satisfactorily explained on any other hypothesis.

In 1815 a chestnut mare, seven-eighths Arabian, belonging to the Earl of Morton, was covered by a quagga (a species of zebra): the hybrid produce resembled the sire in color and in many peculiarities of form.

"In 1817, 1818, and 1821, the same mare was covered by a very fine black Arabian horse, and produced successively three foals, and, although she had not seen the quagga since 1816, they all bore his curious and unequivocal markings."[1]

[1] This remarkable case was first published in the "Philosophical Transactions," 1821, p. 20. It has been repeatedly cited by writers on breeding, some of whom have apparently been misled by making quotations at second hand. A writer in the *Farmer's Magazine* mentions the case of "a thorough-bred mare belonging to Sir Gore Ousely," that was covered by a zebra, and, on the authority of Mr. Blanc, states that "Lord Morton had a mare covered by a quagga—a kind of large ass,"

It is stated, on the authority of Mr. William Good-
win, veterinary surgeon to her Majesty, that "several
of the mares in that establishment" (royal stud at
Hampton Court) "had foals in one year, which were
by Actæon, but which presented exactly the marks of
the horse Colonel, a white hind-fetlock, for instance,
and a white mark or stripe on the face; and Actæon
was perfectly free from white. *The mares had all
bred from Colonel the previous year.*" [1]

"A colt, the property of the Earl of Suffield, got
by Laurel, so resembled another horse (Camel) that it
was whispered, nay, even asserted, at Newmarket, that
he must have been got by Camel. It was ascertained,
however, that the only relation which the colt bore to
Camel was, that the latter had served his mother the
previous season." [2]

Mr. George T. Allman, of Tennessee, gives the
following case, that came under his own observation:
"I bred a bay mare, black points, to Watson, a son of
Lexington, who is a golden chestnut, large star, both

the results in each case being the same. As the mare belonging to
Lord Morton, that was covered by a quagga, was afterward sent to Sir
Gore Ousely, and produced colts by a black Arabian horse, the two
cases are readily resolved into one. (*See* Darwin's "Animals and Plants
under Domestication," vol. i., p. 484; *Farmer's Magazine*, vol. xxxv., p.
130; Walker on "Intermarriage," p. 244; "Principles of Breeding,"
by Goodale, p. 46; *British and Foreign Medico-Chirurgical Review*,
July, 1863, p. 183.)

[1] *Farmer's Magazine*, vol. xxxv., p. 130. *See* also "Principles of
Breeding," by Goodale, p. 47; *Journal of the Highland Agricultural
Society*, 1857–'59, p. 26.

[2] *Journal of the Highland Agricultural Society*, 1857–'59, p. 26.
See also "Principles of Breeding," by Goodale, p. 47; *Farmer's Magazine*, vol. xxxv., p. 130.

hind and near front ankles white. After dropping her foal to Watson, I bred the same mare to my saddle-stallion, Prince Pulaski, a very dark chestnut, no white save a very small star; this produce was a *fac simile* of Watson *in every particular.*"[1]

"Alexander Morrison, Esq., of Bognie, had a fine Clydesdale mare which, in 1843, was served by a Spanish ass and produced a mule. She afterward had a colt by a horse, which bore a very marked likeness to a mule—seen at a distance, every one set it down at once as a mule. The ears are nine and a half inches long, the girth not quite six feet, and stands above sixteen hands high. The hoofs are so long and narrow that there is a difficulty in shoeing them, and the tail is thin and scanty. He is a beast of indomitable energy and durability, and is highly prized by his owner."[2]

A similar case is recorded by Dr. Burgess, of Dedham, Massachusetts, who says, "From a mare which had once been served by a jack, I have seen a colt so long-eared, sharp-backed, and rat-tailed, that I stopped a second time to see if he were not a mule."[3]

Dr. H. B. Shank, of Lansing, Michigan, informs me that a mare belonging to himself having produced a mule, was afterward bred to a Morgan stallion with remarkably fine ears; the ears of the colt were large and coarse, presenting a close resemblance to those of a mule. A second colt produced by the

---

[1] *Rural Sun*, as quoted in *National Live-Stock Journal*, June, 1877, p. 245.

[2] " Principles of Breeding," by Goodale, p. 48.

[3] *Country Gentleman*, 1870, p. 426.

mare to the same stallion had the head and ears of its sire.

A repetition of the procreative function was apparently necessary, on the part of the stallion, to overcome the transmitted influence of the jack remaining from a former impregnation. Similar cases have frequently been observed by persons engaged in mulebreeding.

"A pure Aberdeenshire heifer was served with a pure Teeswater bull, by which she had a first-cross calf. The following season the same cow was served with a pure Aberdeenshire bull; the produce was a cross-calf, which, when two years old, had very long horns, the parents being both polled.

"Again, a pure Aberdeenshire cow was served, in 1845, with a cross-bull—that is to say, an animal produced between a first-cross cow and a pure Teeswater bull. To this bull she had a cross-calf. Next season she was served with a pure Ayrshire [Aberdeenshire?] bull; the produce was quite a cross in shape and color."[1]

Mr. Shaw, of Leochel-Cushnie, "put six pure-horned and black-faced sheep to a white-faced hornless Leicester ram, and others of his flock to a dun-faced Down ram. The produce were crosses between the two. In the following year they were put to a ram of their own breed, also pure. All the lambs were hornless and had brown faces. Another year he again put them to a pure-bred horned and black-faced ram. There was a smaller proportion this year impure; but two of the produce were polled, one dun-

---

[1] Quoted from Dr. Harvey's paper on "Cross-Breeding," in the *Journal of the Highland Agricultural Society*, 1857-'59, p. 26.

faced, with very small horns, and three were white-faced—showing the partial influence of the cross even to the third year." [1]

"A small flock of ewes belonging to Dr. W. Wells, in the island of Grenada, were served by a ram procured for the purpose—the ewes were all white and woolly; the ram was quite different—of a chocolate color, and hairy, like a goat. The progeny were of course crosses, but bore a strong resemblance to the male parent. The next season Dr. Wells obtained a ram of precisely the same breed as the ewes, but the progeny showed distinct marks of resemblance to the former ram in color and covering.

"The same thing occurred on neighboring estates under like circumstances." [2]

Mr. Darwin cites the following case from the "Philosophical Transactions," 1821: "Mr. Giles put a sow of Lord Western's black-and-white Essex breed to a wild-boar of a deep chestnut-color, and the 'pigs produced partook in appearance of both boar and sow, but in some the chestnut-color of the boar strongly prevailed.' After the boar had long been dead the sow was put to a boar of her own black-and-white breed—a kind which is well known to breed very true, and never to show any chestnut-color—yet from this union the sow produced some young pigs which were plainly marked with the same chestnut-tint as in the first litter." [3]

F. Sherman, of Ash Grove Farm, Fairfax County,

[1] *Farmer's Magazine*, vol. xxxv., p. 130.
[2] "Principles of Breeding," by Goodale, p. 49.
[3] "Animals and Plants under Domestication," vol. i., p. 485.

Virginia, relates his experience as follows: " Three years ago one of my Essex gilts was served by a little sandy 'scrub' boar that slipped through the fence from the road-side. . . . The result was a litter of four pigs, two pure black, the others sandy, with black spots. In due time she was again served by one of my Essex boars. Two pigs of the resulting litter were again sandy, with black spots. . . . To give this instance its just weight as evidence on the point in question, it is proper to state that the sow, and the boar by which she was served the second time, were pedigree animals of undoubted purity and excellent descent; that no pigs except thorough-bred Essex are kept on the farm for any purpose, and that sows brought here for service by boars are not allowed to run with my animals. After getting one litter of half-bloods, thorough precautions were taken to prevent a repetition of the mishap." [1]

Two similar cases have come under my own observation, under circumstances that do not admit of doubt as to the parentage of the offspring that inherited a stain through a previous impregnation of their dam. Several years ago a Chester white sow, belonging to the Michigan State Agricultural College, had a litter of cross-bred pigs by an Essex boar. The pigs were all more or less spotted with black, but in several of them the white predominated.

The next season the same sow had pigs by a pure Suffolk boar, but they all had black spots, and some of them were more than one-half black. One remarkable feature of this case was the peculiar distribution

[1] *Country Gentleman*, 1877, p. 462.

of the color in several instances. In some the front half of the body was white and the back half of the body black, while in others the colors were reversed, the front half of the body being black and the back half white. The line of demarkation between the black and the white was so regular and well-defined that, if it had been possible to divide the two animals transversely on the line between the white and the black, and transpose the parts before putting together again, a purely white pig and a purely black pig might have been made from the two that were half black and half white.

It is worthy of notice in this connection that E. W. Cottrell, in speaking of a cross of the Suffolk and Essex swine says: "One peculiar feature with the color of this cross is, that invariably the black is in excess upon the hind-part of the animal, while the white predominates upon its fore-parts. I have seen them one half pure black and the other half pure white, with the dividing line where the colors meet forming a circle around the body at the middle." [1]

In July, 1877, in company with my friend Dr. H. B. Shank, of Lansing, Michigan, I visited the farm of Mr. A. N. Gillett, in the town of Delta, Ingham County, where we saw a litter of pigs out of a pure Berkshire sow, and got by a pure Berkshire boar.

More than one-half of the pigs were apparently Poland-China in the form of the head, and their bodies were spotted with sandy-white. We were informed by Mr. Gillett that the preceding year the

---

[1] *Michigan Farmer*, as quoted in William Smith's "Catalogue of Breeding Swine," p. 32.

dam of these pigs had produced a litter of pigs, by a
Poland-China boar, that were marked in the same
manner with sandy-white spots. The sow was bred
under my direction, at the Michigan Agricultural
College, three years ago, and the stock from which
she was descended had not shown any variations from
the pure Berkshire type.

Mr. George T. Allman, of Tennessee, in the paper
noticed below, says: " I bought a trio of Neapolitan
hogs, a boar and two sows; I first bred a very fine,
pure-bred Berkshire sow to the Neapolitan boar; after
farrowing, I bred her to Toronto Chief, a Berkshire
boar, bred by Bush Brothers, Clark County, Kentucky
(from an imported pair), and every time the sow far-
rowed, up to her death, she produced pigs with little
or no hair, like the Neapolitan."[1]

Mr. Darwin, on the authority of Dr. Bowerbank,
gives the following striking case: "A black, hairless,
Barbary bitch, was first impregnated by a mongrel
spaniel, with long brown hair, and she produced five
puppies, three of which were hairless and two covered
with *short* brown hair. The next time she was put
to a full black, hairless, Barbary dog; but the mis-
chief had been implanted in the mother, and again
about half the litter looked like pure Barbarys, and
the other half like the short-haired progeny of the
first father."[2]

The following case is given by Mr. George T. Allman, of Tennessee : "I bought at ' Woodburn,' Kentucky, the shepherd-dog York, from the pair the late

[1] *Loc. cit.*, p. 245.
[2] "Animals and Plants under Domestication," vol. i., p. iii.

R. A. Alexander imported. (York was got by the famous Spring.) At the same time I bought a bitch, Fannie (Scotch collie), first produce of the imported pair owned by R. A. Alexander. Fannie and York were the only dogs on the farm, and are both still living. Fannie came in heat three different times, was put in a stall and secured from any intrusion, but she would not allow York to serve her. The third time she came in heat a young man, who was out hunting near my place with a liver-and-white colored pointer, suggested that I let his dog into the kennel. I did so, and he served Fannie, and afterward the shepherd-dog York did also. Half of the litter of pups were colored *precisely* as the pointer, and the remainder were about equally divided in color, part taking after York and part after Fannie. Since then Fannie has been coupled *only with pure* shepherd-dogs, yet *every* litter of pups has from one to two marked *precisely* like the pointer that first served her." [1]

The same paper contains the two following cases, given by G. A. Baxter, M. D., of Chattanooga, Georgia: "Colonel L——, of Chattanooga, had a white English bull-bitch, which by chance took a dog of different species. Though he ever afterward tried to preserve the white breed pure from her, she continued until her death, with every litter, to bear one or two yellow pups. Some of the pups I have seen myself, and he yet owns one in Chattanooga."

"Mr. C——, of Chattanooga, has a small-sized, bluish-tinted shepherd, of a peculiar breed, and im-

[1] *Rural Sun*, as quoted in *National Live-Stock Journal*, 1877, p. 245.

ported, I think, having very straight hair. Three years ago this bitch was bred by him to another shepherd of a different species—a large, shaggy-haired breed. I saw her last litter of pups, after she had been confined during the whole period of heat with a dog of her own species, and, without knowing the fact of her having been so bred, remarked upon the singular difference in size, shape, and appearance, of two of the litter from the remainder—they being half as large again, and seemingly of another breed entirely from her or the father. I was told then of the above-mentioned facts, which explained conclusively the result, and I think logically and truly." [1]

Prof. Agassiz states that he had "experimented with a Newfoundland bitch, by coupling her with a water-dog, and the progeny were partly water-dog, partly Newfoundland, and the remainder a mixture of both. Future connections of the same bitch with a greyhound produced a similar litter, with hardly a trace of the greyhound. He had bred rabbits with the laws established by this experiment, and had at last so impregnated a white rabbit with the gray rabbit that connection of this white rabbit with a black male invariably produced gray." [2]

A celebrated breeder of Short-Horns, of my acquaintance, bred the females of a light-colored family to a red bull, and afterward to a bull of their own family; and he succeeded, in this manner, in producing the desired shades of color in the offspring of the light-colored females.

[1] *National Live-Stock Journal*, 1877, p. 245.
[2] " Agricultural Report of Massachusetts," 1863, p. 57.

The same influence has been observed in the human family. "A woman may have, by a second husband, children who resemble a former husband, and this is particularly well marked in certain instances by the color of the hair and eyes.

"A white woman, who has had children by a negro, may subsequently bear children to a white man, these children presenting some of the unmistakable peculiarities of the negro race."[1]

Several theories have been advanced to explain the manner in which this peculiar influence has been transmitted.

As the first cases that attracted the attention of physiologists were observed among mammals, it was supposed that the mother was impressed with the paternal characteristics of the fœtus during its intrauterine existence.

In his remarks on this subject, Dr. Carpenter says: "Some of these cases appear referable to the strong mental impression left by the first male parent upon the female; but there are others which seem to render it more likely that the blood of the female has imbibed from that of the fœtus, through the placental circulation, some of the attributes which the latter has

---

[1] "Physiology of Man," by Flint, vol. v., p. 347. *See* also "Human Physiology," by Carpenter, p. 970; "Cyclopædia of Anatomy and Physiology," vol. iv., pp. 1341–1365; *Journal of the Royal Agricultural Society*, vol. xvi., p. 23; *British and Foreign Medico-Chirurgical Review*, July, 1863, p. 183. Additional references are made by Darwin to cases of this kind of influence that I have not an opportunity to consult, as follows: Broun, in his "Geschichte du Natur," 1843, B. 11, S. 127; and Martin's "History of the Dog," 1845, p. 104; "Animals and Plants under Domestication," vol. i., p. 485, note.

derived from its male parent, and that the female may communicate these, with those proper to herself, to the subsequent offspring of a different male parentage."[1]

Mr. James McGillivray, a veterinary surgeon of Huntly, presents essentially the same theory, as he believes that, when a female of any pure breed has been impregnated by a male of another breed, she becomes a cross, " the purity of her blood being lost in consequence of her connection with the foreign animal."[2]

Dr. Harvey, who had advocated the same theory, afterward observes : " Since then I have learned that many among the agricultural body in this district are familiar, to a degree that is annoying to them, with the facts there adduced in illustration of it—finding that after breeding crosses their cows, though served with bulls of their own breed, yield crosses still, or rather mongrels ; that they were already impressed with the idea of contamination of blood as the cause of the phenomenon ; that the doctrine so intuitively commended itself to their minds, as soon as stated, that they fancied they were told nothing but what they knew before."[3]

If the influence of the male upon the offspring of the same mother by another male were limited to the class of mammals, this theory might be accepted as a plausible explanation of the cases that have been presented ; but there are instances in which a similar

---

[1] " Human Physiology," p. 970.

[2] " Principles of Breeding," p. 52.

[3] *Edinburgh Journal of Medical Science*, 1849, as quoted by Goodale, *loc. cit.*, p. 53.

influence has been observed in fowls—where the egg
is separated from the mother before the embryo is
developed—that cannot have been produced by a contamination of the blood of the mother by that of the
embryo. But another theory that has been advanced
to explain the manner in which this influence is transmitted in mammals must be noticed.

Prof. James Law, after mentioning some of the
theories that had been advanced to explain the phenomena under discussion, says : " But a simpler and
more satisfactory explanation may be found. It is a
well-known pathological fact that adjacent cells tend
to ingraft their plastic or formative powers upon each
other. I prick my skin with a needle. Immediately
the injured cells and nuclei undergo a rapid increase
in size and numbers, but the effect does not end there ;
those adjacent take on a similar action, and the extent
of the resulting inflammation is only limited by that
of the injury and the susceptibility of the parts.
Again, in placing a slice of epidermis in the middle
of a raw sore we inoculate the cells of the adjoining
granulations, and empower them to develop epidermic
structure. How, then, can we avoid the conclusion
that the impregnated ovum impresses its own characters on the mass of the decidua, and, through this, on
the maternal placenta, and that this in turn impresses
its characters on the decidua and embryo of the next
succeeding generation ?" [1]

This theory is certainly an ingenious one, but it
does not furnish a satisfactory explanation of all of

[1] "Reports and Papers of the American Public Health Association,"
vol. ii., p. 253.

the observed cases of this peculiar influence; and it also fails to take into account certain physiological facts that are difficult to reconcile with it.   It is well known that the placenta and decidua are temporary organs that disappear at the time of parturition, and that even the mucous membrane itself is removed and replaced with new tissue.

Dr. Dalton says : "Another very remarkable phenomenon connected with pregnancy and parturition is the appearance in the uterus of a *new mucous membrane*, growing underneath the old, and ready to take the place of the latter after its discharge.   If the internal surface of the body of the uterus be examined immediately after parturition, it will be seen that at the spot where the placenta was attached every trace of mucous membrane has disappeared.

" The muscular fibres of the uterus are here perfectly exposed and bare, while the mouths of the ruptured uterine sinus are also visible, with their thin, ragged edges hanging into the cavity of the uterus, and their orifices plugged with more or less abundant bloody coagula.   Over the rest of the uterine surface the decidua vera has also disappeared.   Here, however, notwithstanding the loss of the original mucous membrane, the muscular fibres are not perfectly bare, but are covered with a thin, semitransparent film, of a whitish color and soft consistency.

" This film is an imperfect mucous membrane, of a new formation, which begins to be produced underneath the old decidua vera as early as the beginning of the eighth month." [1]

[1] "Human Physiology," pp. 621, 622.

"At birth," says Dr. Marshall, "the embryonal vascular portion of these membranes, whether it be a diffused, cotyledonous, zonular, or discoidal placenta, is always detached. In the case of the zonular and discoidal forms of the placenta, where a true decidua is developed, *a part of the maternal tissues is also* separated at the same time.

"Where there is no decidua, as in the diffuse and cotyledonous forms, the fœtal villi are merely detached from the surfaces or recesses into which they fit. In the latter cases parts of the maternal tissues, especially of the veins and venous lacunæ, come away."[1]

It does not seem probable that an impression received by these temporary structures should be transmitted, through their influence, to subsequent impregnations.

The numerous instances of the influence of a previous impregnation upon offspring by another male that have been observed in fowls, to which we now direct our attention, must, however, be *fatal* to this theory, as well as that of blood contamination.

Mr. W. H. Smith, of Lexington, Kentucky, makes the following statement: " On or about the first day of February, 1873, I loaned a prime Dark Brahma cock, that was a good, vigorous bird, to Mr. James Fought, of this city. He put him with a lot of Light Brahma hens, with which a Houdan cock had been running previously. The hens laid, set, hatched, and

---

[1] "Outlines of Physiology," p. 960. *See* also "Text-Book of Human Physiology," by Flint, p. 943; Flint's "Physiology of Man," vol. v., pp. 376, 454; "Cyclopædia of Anatomy and Physiology," vol. v., p. 659; Carpenter's "Human Physiology," p. 980.

raised their chicks, laid and hatched again, and the second litter of chicks had the Houdan marks. There was no Houdan blood in the Light Brahma hens, neither was there any other cock with the hens from the time he got the Dark Brahma cock."[1]

Mr. A. W. Frizzell, of Baltimore County, Maryland, makes the following statement: "I once purchased a trio of pure-bred Dark Brahma fowls from a breeder of no small note, and a trustworthy man (I speak from experience, for I was once employed by this gentleman, and do know him to be trustworthy), which fowls had taken the first premium at the Carroll County (Kentucky) Fair in 1871. I brought those fowls home, and in the yard was also a Light Brahma cock, which I did not dispose of for some time, and in the mean time he was mating with these dark hens; any effects of this I thought would soon run out. After a while I disposed of the light cock, and kept none but the dark one, or had none nearer than a mile. Nevertheless, three years afterward I see those light, or half-light, chicks coming from those two hens."[2]

"A Mr. Payne, in England, had two Spanish pullets running with both a Spanish and Cochin cock. After they began to lay the Cochin was removed, and *six weeks* after the eggs were saved and set; but the chickens were feather-legged, in all other points resembling the Spanish.

"On another occasion the same gentleman allowed

[1] *The Poultry World*, as quoted in *The Country Gentleman,* 1873, p. 475.

[2] *The Country Gentleman*, 1877, p. 151.

a Black-red game-hen, which laid while with chickens, to run a few hours with a Brown-red cock, and nine eggs produced chickens, which all resembled the father, or Brown-red.

"Another English gentleman, when residing in Canada, sold his Brahma cock and one hen, allowing the hen left to run afterward with a Spangled Hamburg which had five hens of his own. Every egg laid for ten days produced a pure Brahma chick, that laid on the eleventh day was a half-breed.

"In America, a Mr. Woodward bought in March some Spanish pullets which had been running all the winter with a native cock, and, though no eggs were set till two months after purchase, all the chicks even then showed the native points in a high degree.

"Another gentleman breeding Games, finding a neighbor's feather-legged Bantam cock come over his fence, penned his fowls in securely, and saved no eggs for a month after; but several chicks still had feathered legs, though with no other sign of the cross."[1]

Mr. E. W. Barnes, of Plympton, "allowed a neighbor's Brown Leghorn cock to pass three days among his pen of eight one-year-old Light Brahma pullets, 'for experiment's sake,' he said. The Brown Leghorn cock was removed, and he has never once had anything on his premises since but the Light Brahmas, of both sexes, 'pure.'

"From the eggs set within a week after the Brown Leghorn cock was sent home a third of the chicks, when hatched, came brown, speckled-brown, or patched with brown, that same summer.

[1] "The Illustrated Book of Poultry," by Wright, pp. 129, 130.

"Out of the eight hens he saved four (which were alive a year ago), and last season—two years after the Brown Leghorn cock was dead—more than one-quarter of Mr. Barnes's chicks, bred from the old Light Brahma hens with a Light Brahma cock *only* since, came spotted, speckled, and splashed with *brown* feathers."[1]

Mr. Charles H. Edmonds, of Melrose, "allowed a Sebright cock to run for a few weeks" with his Light Brahma fowls. "In the fall his Light Brahma chicks were marked with distinct Golden Sebright feathers, and for two years succeeding this marking showed itself on scores of his chicks, from this very flock of Light Brahmas, when the Sebright cock had been gone from his premises over two seasons."[2]

In discussing this class of cases, Mr. Wright remarks: "But the fact remains—proved beyond the possibility of doubt—that again and again hens of different breeds, and female animals of various kinds, after the birth of half-bred offspring, have ever afterward manifested a plainly-evident tendency to produce offspring bearing more or less strong traces of the same characters. This tendency greatly varies, and cannot therefore be calculated; but it exists, and tends to show that a given chick may, in a certain mythical sense, have two fathers, or rather that the progeny of one bird is in some mysterious way modified by the previous union with another.

"The most probable explanation is, that as habit is the developed tendency to do again what has already been done, so the female reproductive system,

---

[1] *The Poultry World*, October, 1877, p. 326.
[2] Ibid., p. 327.

having once given birth to offspring having a strongly-marked character, becomes in a degree *moulded* to that character, and tends again to produce it.

"At all events the teaching of this fact is plain, and we would never, on any account, allow any valued hens to mate with another breed. We have known ourselves several cases in which hens once crossed have reproduced strong cases of that cross *two years* afterward; and many otherwise unaccountable occurrences, which have given rise to bitter recriminations, may be thus very easily explained."[1]

The intensity of the influence of the male element of fertilization upon the ova seems to vary widely in different species of animals. In many species a single act of copulation is sufficient to impregnate a number of eggs, while in others a repetition of the act is apparently required to produce fecundation.

Mr. Wright has collected a number of instances showing that the eggs of the hen are fertile from four to sixteen days after separation from the cock;[2] and it is a fact well known to breeders that, with turkeys, a single copulation is sufficient to impregnate all the eggs of one "laying,"[3] while it is stated by Mr. Chapin, of Milford, Massachusetts, "that a hen-turkey would lay two or three successive litters of eggs, having been impregnated only for the first litter."[4]

[1] *Loc. cit.*, p. 130.
[2] Ibid., pp. 129, 130.
[3] *Journal of the Royal Agricultural Society*, vol. xii., p. 198; Tegetmeier's "Poultry-Book," p. 273; "The Illustrated Book of Poultry," by Wright, pp. 130, 519.
[4] "Agricultural Report of Massachusetts," 1863, p. 57.

Agassiz has shown that turtles begin to copulate at the age of seven years, but do not lay until they are eleven years old. They copulate twice each year for four years, before the eggs are fully matured. "Upon opening large numbers of young *Chrysemys picta*,[1] it was ascertained that, up to their seventh year, the ovary contained only eggs of very small size, not distinguishable into sets; but that with every succeeding year there appears in that organ a larger and larger set of eggs, each set made up of the usual average number of eggs which this species lays, so that specimens eleven years old for the first time contain mature eggs, ready to be laid in the spring."

From observations made by Agassiz, "it appears that the first copulation coincides with a new development of the eggs, in consequence of which a certain number of them, equal to that which the species lays, acquire a larger size, and go on growing for four successive years before they are laid, while a new set is started every year, at the period of copulation in the spring, enabling this species to lay annually from five to seven eggs after it has reached its eleventh year."[2]

After a careful examination of all the known facts bearing upon this interesting subject, Agassiz became satisfied that "the first copulation only determines the further growth of a certain number of eggs, which require a series of successive fecundations to undergo their final development;" and that "in turtles a repetition of the act, twice every year for four successive years, is necessary to determine the final development

[1] A common fresh-water turtle.
[2] "Embryology of the Turtle," by Agassiz, pp. 490, 491.

of a new individual, which may be accomplished in other animals by a single copulation."[1]

The repeated fertilization of the eggs of turtles is apparently analogous to the phenomena observed in the transmitted influence of a previous impregnation to the offspring by a subsequent impregnation, as pointed out by Agassiz in a lecture before the Massachusetts Board of Agriculture in 1860; and he remarks that his experiments with dogs, that have already been mentioned, seem to show that "the impregnation of an ovum may take place a long time previous to its development, and that it probably only requires the stimulus of future connections with a male to bring it into existence."[2]

In a subsequent lecture, in speaking of the influence of a previous impregnation upon offspring at a later period, Agassiz says: "It therefore shows what I have satisfied myself to be the truth among other animals, by numerous experiments; that the act of fecundation is not an act which is limited in its effect, but that it is an act which affects the whole system, the sexual system especially, and in the sexual system the ovary to be impregnated hereafter is so modified by the first act that later impregnations do not efface that first impression."[3]

This is undoubtedly the most rational explanation of the cases under consideration that has been presented, and there are additional facts which show that the male element of fertilization may extend its influ-

---

[1] *Loc. cit.*, p. 491.
[2] "Agricultural Report of Massachusetts," 1863, pp. 56, 57.
[3] Ibid., 1856-'57, p. 84.

ence to the ovary itself, as well as to the germs that are not fully developed.

During the period of heat in the lower animals, and of menstruation in women, one or more germs are matured and escape from the ovary, so that the term periodical ovulation has been used to designate the process.

When a germ is thus liberated the walls of the follicle that contained it become thickened, and a peculiar cicatrix is formed, which is called the *corpus luteum*. If impregnation of the germ has not taken place, the *corpus luteum* attains its maximum of development at the end of three weeks (measuring three-fourths of an inch in length and one-half inch wide), and then gradually diminishes in size, so that a minute cicatrix remains at the end of seven or eight weeks, and in the course of seven or eight months it entirely disappears.

When the germ is impregnated the *corpus luteum* attains a greater development, continuing its growth to the end of the fourth month, when it measures seven-eighths of an inch in length, and three-fourths of an inch in depth.

During the fifth and sixth months it remains unchanged, but diminishes again during the seventh, eighth, and ninth months, when it measures half an inch in length and three-eighths of an inch in depth. Several months after delivery the *corpus luteum* entirely disappears.[1]

The mere fact of impregnation seems to determine

[1] Dalton's "Human Physiology," pp. 564–573; Flint's "Physiology of Man," vol. v. ("Generation"), pp. 307–312.

the greater or less development and duration of the *corpus luteum*, and, although it has been supposed that this difference is owing to the greater vascular activity of the generative organs of the female during pregnancy, it appears probable, from the facts that have been presented, that the *corpus luteum* of pregnancy derives its distinctive peculiarities from the direct influence of the male element upon the ovary.

Mr. Darwin cites a number of instances in the vegetable kingdom to show the " direct action of the male element on the mother-form," and he comes to the conclusion that " the male element not only affects, in accordance with its proper function, the germ, but the surrounding tissues of the mother-plant." '

After citing some of the cases that have already been presented of the influence upon offspring of a previous impregnation of the mother, Mr. Darwin says, " The analogy from the direct action of foreign pollen on the ovarium and seed-coats of the mother-plant strongly supports the belief that the male element acts directly on the reproductive organs of the female, wonderful as is this action, and not through the intervention of the crossed embryo." '

It will be observed that this explanation of the continued influence of the male upon offspring by another male is precisely the same as that given by Agassiz, and it is believed that there is a strong preponderance of evidence in its favor.

In the first observed cases, it was claimed that this peculiar influence of the male was limited to the first

---

[1] "Animals and Plants under Domestication," vol. i., p. 483.
[2] Ibid., p. 486.

impregnation of the female only, but there is good reason to believe that every impregnation may leave its impress upon partly-developed germs, and be thus transmitted with the characters of a subsequent fecundation.

The intensity of the influence of the male may be impaired by an excessive use of the procreative organs, and it has been observed in fowls that when the male is " over-mated " the eggs are sometimes imperfectly impregnated.

Mr. Wright remarks that " it is a notorious fact that when a cock is over-mated the eggs always hatch in a very unsatisfactory manner ; " and he adds: " But besides mere fertility there are other considerations ; and, in the first place, it appears indisputable that eggs may be so far fertilized as to commence hatching, and yet not have sufficient vigor to complete the process successfully. The number of cases where such experiments have been made as we have quoted, in which part of the eggs produced showed signs of hatching but *did not hatch*, is proportionately very great, and the conclusion will not be lost on the intelligent breeder.

" But still further, and coming back to the considerations with which we commenced this part of the subject, it is utterly impossible to resist the conclusion that, beyond fertilization, the act of union exerts, in many cases, a more mysterious and far-reaching influence." [1]

The same writer is inclined to the belief that, when a hen-turkey is mated but once for an entire laying of

[1] " The Illustrated Poultry-Book," pp. 130, 131.

eggs, the young birds are not so strong and vigorous as when the male runs permanently with a dozen or fifteen hens.[1]

It seems to be quite generally acknowledged by poultry-breeders that, to produce strong, vigorous off-spring, the cocks should not be allowed to mate with more than from four to six hens, if in confinement, or with twice that number, under the most favorable circumstances, when running at large.[2]

The effects of an impaired influence of the male, in the process of procreation, upon his offspring, need to be more fully investigated; but there are many facts that indicate that this is in all probability a potent cause of degeneracy.[3]

Closely connected with the facts under discussion are the observations that have been made on the lower animals, showing that at least several spermatozoöns (the active male elements of fertilization) are neces-sary to produce a complete impregnation of the germ; but it is perhaps impossible, from the differ-ent conditions presented, to determine experimentally

---

[1] *Loc. cit.*, p. 519.

[2] "The Illustrated Poultry-Book," by Wright, pp. 44, 306; *Journal of the Royal Agricultural Society*, vol. xii., p. 180; "Domestic Fowl," by Richardson, p. 50; Geyelin's "Poultry-Breeding," p. 24; Mowbray on "Poultry," p. 33.

[3] Nordhoff describes the Mormon children as "undersized, loosely built, flabby. . . . The young girls were pale, and had unwholesome, waxy complexions; the young men were small and thin, and looked weak;" but this he attributes to "the hard struggle with life while these youth were babes" (Nordhoff's "California," pp. 42, 43).

(The facts cited above, however, seem to indicate that these peculi-arities may, with greater reason, be attributed to polygamy.)

the same class of facts in animals more highly organized.[1]

" It appears, from Mr. Newport's ingenious experiments, that the contact of a single spermatozoön is not adequate to produce complete fecundation, but that the penetration of a certain number of spermatozoa is requisite; and he has ascertained that fecundation may be effected *partially* (so as to occasion *some*, though not *all*, of the normal changes in the ovum) by a smaller amount." [2]

The last-mentioned fact, it will be noticed, is in accordance with the experience of breeders of fowls, to which reference has already been made.

[1] "Cyclopædia of Anatomy and Physiology," vol. ii., p. 464; Flint's "Physiology of Man," vol. v., p. 353.

[2] Carpenter's "Comparative Physiology," p. 532.

CHAPTER XIII.

THE abnormal peculiarities occasionally observed in animals at the time of birth, that are not recognized as family characteristics, have been popularly attributed to some mysterious influence of the imagination of the mother in the process of intra-uterine development.

This influence is supposed by many to be exerted not only in mammals, where the most intimate relations are known to exist between the mother and the embryo during the period of utero-gestation, but also in fowls, where the egg is separated from the mother before the slightest indications of embryological development can be detected.[1]

The following cases, which have been reported as illustrations of this influence, will be sufficient to show the kind of evidence on which it rests, and the varied results it is claimed to produce :

[1] "Cyclopædia of Anatomy and Physiology," vol. ii., p. 475 ; Wright's " Book of Poultry," pp. 131, 314.

It has even been assumed that birds, in the process of incubation, exert an influence upon the eggs they are hatching that is sufficient to modify the characters of the progeny. In artificial incubation, however, and when the eggs of one species are hatched by another, the inherited characters are not modified.

13

"It is stated that the ambition, courage, and military skill of Napoleon Bonaparte had their foundation in the circumstance that the emperor's mother followed her husband in his campaigns, and was subjected to all the dangers of a military life; while, on the other hand, the murder of David Rizzio in the presence of Queen Mary was the death-blow to the personal courage of King James I., and occasioned that strong dislike of edged weapons for which that crafty and pedantic monarch was said to be remarkable."[1]

At the siege of Landau, in 1793, "in addition to a violent cannonading, which kept the women for some time in a constant state of alarm, the arsenal blew up with a terrific explosion, which few could hear with unshaken nerves. Out of ninety-two children born in that district within a few months afterward, Baron Percy states that sixteen died at the instant of birth; thirty-three languished for from eight to ten months and then died; eight became idiotic, and died before the age of five years; and two came into the world with numerous fractures of the bones and limbs, caused by the cannonading and explosion. Here, then, is a total of fifty-nine children out of ninety-two, or within a trifle of two out of every three, actually killed through the medium of the mother's alarm, and the natural consequences upon her own organization."[2]

Mr. Boswell relates the following, on the authority of "Mr. Mustard, an extensive farmer on Sir James

---

[1] "Cyclopædia of Anatomy and Physiology," vol. ii., p. 474.
[2] Carpenter's "Human Physiology," p. 1011.

Carnegie's estate in Angus," Scotland: "One of his cows chanced to come into season while pasturing on a field which was bounded by that of one of his neighbors, out of which field an ox jumped and went with the cow until she was brought home to the bull.

"The ox was white, with black spots, and horned. Mr. Mustard had not a horned beast in his possession, nor one with any white on it. Nevertheless, the produce of the following spring was a black-and-white calf, with horns." [1]

"It is related that, at the time when a stallion was about to cover a mare, the stallion's pale color was objected to, whereupon the groom, knowing in the effect of color upon horses' imaginations, presented before the stallion a mare of a pleasing color, which had the desired effect of determining a dark color in the off-spring. This is said to have been repeated with success in the same horse more than once." [2]

"Prof. Dalton, whose accuracy upon such a point cannot be questioned, noted the following: While he was lecturing upon the subject of generation, at the College of Physicians and Surgeons of New York, the janitor of the college called his attention to his child, which presented a deformity of the external ear, as though a portion had been taken off with a sharp instrument. The janitor stated that his wife, during her pregnancy, dreamed that she saw a man with a similar deformity. This dream was very vivid, and she immediately related it to her husband. They

---

[1] "Transactions of the Highland Agricultural Society," vol. i., p. 28.
[2] "Cyclopædia of Anatomy and Physiology," vol. ii., p. 474.

both believed that this was the cause of the deformity of the child."[1]

A gentleman of my acquaintance states that he saw a lamb that resembled a rabbit in the form of its feet; the dam had been kept in a pasture where rabbits were numerous.

"A woman whose children had previously been healthy, six weeks before conception is suddenly frighted by a beggar who presents a stumped arm and a wooden leg, and threatened to embrace her; the next child had only one stump leg and two stump arms."

"A young woman, frighted in her first pregnancy by the sight of a child with hare-lip, bears a child with a complete deformity of the same kind; her second child had merely a deep slit, and her third no more than a mark in the same place."

"A child is born with a hare-lip, which was caused by the mother's frequently seeing a child with the same deformity during her pregnancy."

"A lady in London, who is frightened by a beggar presenting the stump of an arm to her, bears a child wanting a hand."

"A child is born covered with hairs, in consequence of the mother having been in the habit of beholding a picture of St. John the Baptist."

"A woman gives birth to a child covered with hair and having the claws of a bear, from her constantly beholding the images and pictures of bears hung up everywhere in the dwelling of the Ursini family, to which she belonged."[2]

---

[1] Flint's "Physiology of Man," vol. v., p. 351.

[2] The last six cases are quoted from Dr. Allen Thomson, who

The most remarkable case of supposed influence of the imagination that has come to my knowledge was communicated to me by Mr. John B. Poyntz, a breeder of Jersey cattle, Maysville, Kentucky, in a letter dated December 18, 1872. At my request he made a more particular statement of the attending circumstances, substantiated by affidavits, which was published in *The Bulletin*, Maysville, Kentucky, February 18, 1875.

The published statement of Mr. Poyntz was as follows : "Alderney Farm, near Maysville, Kentucky, January 18, 1875. In the year 1863 the theory of Prof. Thury, of Geneva, Switzerland—the production of sex at will—was undergoing investigation on my farm. For that purpose I selected a lot of Alderney heifers and a bull; none of them were marked or branded, nor were their ancestors subsequent to 1850. In the month of July the cattle were placed on a woodland pasture, well provided with water and blue grass, and in the pasture were placed a number of government horses, where they remained several weeks. Each and every horse was branded on the lower part of the left shoulder with the letters U. S. In the spring and summer of 1864 the heifers had calves. One of the number produced a fawn-colored

copies over forty similar cases from Burdach and Dr. Blundell ("Cyclopædia of Anatomy and Physiology," article "Generation," vol. ii., p. 475).

In connection with the last case, Dr. Thomson remarks that "it is not stated by the author of 'Waverley' whether anything of the kind ever happened in the Bradwardine family;" and he might with equal propriety have raised the question as to the frequency of such malformations among the inhabitants of Berne.

or reddish calf, and on the lower part of the shoulder were the letters U. S., formed of white hairs, plainly to be seen by casual observers; was shown by me to friends and visitors; and in due time my U. S. heifer had a calf which was marked with U. S. in the same place as her dam; the letter S. was not so perfectly formed as on the dam, but was too plain to be taken for anything other than the letter S. In the growth of these cattle or cows the letters moved higher upon the shoulder and appeared to elongate, and, in five or six years, the character or form of the letters was lost and appeared only as numerous small white specks or spots. This is the statement in full, which I propose to substantiate by the statement of others sworn to before the proper authorities of this county.

" JOHN B. POYNTZ."

" This day appeared John B. Poyntz, who is well known to me, signed the above statements, and made oath that they were true.

[SEAL.]   " C. B. PIERCE, *Notary Public.*" [1]

---

[1] Accompanying this statement were the following affidavits:

"STATE OF KENTUCKY, } &t.
MASON COUNTY, }

" F. II. Bierbower, a resident of Maysville, in the State above written, being first duly sworn, states that in the summer of 1863, while he was Captain of Company A, of Fortieth Kentucky Mounted Infantry, he pastured some twenty or thirty head of horses on the farm of John B. Poyntz, near the city of Maysville; the said horses were the property of the United States, and were distinctly branded on the left fore-shoulder with the letters U. S.; the affiant further states that cattle were confined at the same time with said horses in the same pasture.

" F. II. BIERBOWER."

This case might be cited to show the literary ability and patriotism of the Alderneys, as well as their powers of observation and active imagination.

The longing of the mother for strawberries, grapes, cherries, etc., has been supposed to produce marks on the offspring that present a fancied resemblance to the object of desire.

The cases in which the habitual condition of the mother is repeated in her offspring do not differ in

"This day appeared F. H. Bierbower, who is well known to me, signed the above statements, and made oath that they were true.

[SEAL.] "CHAS. B. PIERCE, *Notary Public*."

"MAYSVILLE, KENTUCKY, *January* 21, 1875.

"I hired on the Alderney farm of John B. Poyntz, and had the care of his herd of cattle, and remember well the circumstance of the Government horses being pastured with the cattle, also the birth of the calf marked with the letters U. S., in white hairs on the shoulder, the calf being of reddish or fawn color, and that when she had a calf it was marked in the same place and with the same letters as the dam—the letter S. was not so perfectly formed, but could not be mistaken for any other letter. SAMUEL OLDHAM."

"This day appeared Samuel Oldham, who is well known to me, signed the above statements, and made oath that they were true.

[SEAL.] "CHAS. B. PIERCE, *Notary Public*."

"MAYSVILLE, KENTUCKY, *January* 21, 1875.

"I purchased a farm adjoining that of John B. Poyntz, upon which I have lived up to this date, have often seen his U. S. heifer, as she was called, and noticed the letters on her shoulder in white hairs; also remember of her having a calf marked in the same manner on the shoulder; they were shown by Poyntz to his friends and visitors as curiosities. JOHN H. WILSON."

"This day appeared John H. Wilson, who is well known to me, signed the above statements, and made oath that they were true.

[SEAL.] "CHAS. B. PIERCE, *Notary Public*."

any essential particular from the instances of hereditary transmission of acquired habits, that have been
noticed in another chapter.

A habit of the mind of either parent may be transmitted to their progeny in accordance with the same
laws that determine the transmission of any other
character or quality, and it seems to be entirely unnecessary to assume that the imagination of the mother
is an active agent in determining the result.

Malformations of the fœtus have not been attributed by physiologists to the direct influence of the
imagination of the mother, for the following reasons:

" 1. Malformations seldom, or perhaps never, agree
with the apprehensions or fears, *a priori*, of pregnant
women. On the contrary, it often happens that a
woman who has once procreated a malformation and
is continually troubled by the fear of another similar
occurrence, may become the happy mother of a second
well-formed child."

" 2. Malformations occur likewise among the inferior animals—insects, testaceous animals, echinoder-
mata—in which the development of psychical life is
very imperfect, and the oviparous generation of which
must preserve the young from the influence of disor-
dered maternal imagination."

" In the case of twins, as the acephali specially
show, one child may be malformed and the other in
perfect condition, notwithstanding they were both
exposed to the same influence.

" That more deeply-situated organs, the very existence of which may be unknown to the pregnant

woman, may be malformed, as for instance the heart, the intestinal tube, etc."[1]

In most cases of malformation the mental impression that is assigned as a cause is not presumed to have been injurious until the malformation is observed, while the violent shocks that give rise to apprehensions of injury are usually found to have made no impression upon the development of the fœtus.

The anatomical relations of the embryo and its uterine envelopes likewise render it improbable that any mental impressions of the mother can be transmitted to any particular part of the fœtus, to exert a specific influence in its development.

There are many considerations that seem to indicate that malformations of the embryo are determined by fixed organic laws that preclude the intervention of paroxysmal causes.

" We never see in malformed births dissimilar parts fused or united with each other, such as the intestinal tube with the aorta, the arteries with the nerves, etc. Each part, therefore, retains to a certain degree its own independence. . . . The gullet sometimes coalesces with the larynx, and the bladder with the rectum ; but these parts are not originally dissimilar, being developed from a common mass.

" The malformed parts are restricted to their determinate place, according to what Fleischmann denominates *lex topicorum.*

" No malformed organ loses entirely its own char-

[1] W. Vrolik, " Cyclopædia of Anatomy and Physiology," article " Teratology," vol. iv., p. 943.

acter, and no malformed animal loses its generic distinction.   It is therefore justly observed by Söm-
mering that Nature does not deviate *ad infinitum*,
and that even in monstrosities a distinct gradation
and natural order are observable.

" This order appears even—1. *In the number* in
which they occur within a certain space of time.   In
three thousand births in Paris there occurs about one
monster.

" 2. *In the sex.*   In impeded development the
malformed children are more frequently female, in
some sorts of double monsters, male.

" 3. *In a definite proportion between the species
of animals,* and *the most frequent monstrosities in
them.*   Cyclops,[1] for instance, especially with a snout,
occur most frequently in swine ; double monsters in
man.

" 4. *In the constant form of monsters, even among
the most heterogeneous animals.*   Cyclopia, double
monsters, acrania,[2] have in birds precisely the same
characters as in the mammalia.

" 5. *In the greater predisposition to monstrosity
among some animals.*   This is greater among domestic than among wild animals ; greater among the more
perfect than among the less perfect ; three-fourths of
the monstrosities occur among the mammalia, one-
fourth among birds.   They happen seldom among
reptilia, still less frequently among fishes, mollusca,
articulata, and radiata."

" From these premises the consequence is easily

[1] Monsters with one eye.
[2] Headless monsters.

derived that monstrosities do not take place by chance, and therefore do not by any means deserve the so very general appellation of caprices of Nature (*lusus naturæ*). The result of this is, that they often present a quantitative antithesis, according to what Geoffroy St.-Hilaire denominates *loi de balancement*.[1] According to this law the excessive development of one part of the body is often connected with checked formation of another. To *anencephalia*,[2] *cyclopia, spina bifida*,[3] are often joined fingers and toes in excessive numbers; to *sireno-melia*,[4] superfluous vertebræ and ribs; and frequently there occur in double monsters malformations of the head. Meckel saw in one instance this antithesis extend itself over different children of one and the same mother. A girl had on each extremity a superfluous digit; one hand of her sister wanted four fingers, being the number of digits which her sister had in excess, reckoning the four extremities together."[5]

The laws of embryological development furnish a satisfactory explanation of the cases in which there is a fancied resemblance of the fœtus to some of the lower animals.

There is a close correspondence between the embryos of all vertebrate animals in the earliest stages of development; "and it is only with the advance of the developmental process that indications successively

[1] Another name for law of correlation.
[2] Brainless monsters.
[3] Fissure of the spinal column.
[4] Monsters without feet.
[5] W. Vrolik, "Cyclopædia of Anatomy and Physiology," article "Tetratology," vol. iv., pp. 945, 946.

present themselves, which enable us to distinguish, one after another, the characters of the order, the family, the genus, the species, the variety, the sex, and the individual—*the more special features progressively evolving themselves out of the more general,* which is the expression of the law of development, common to all organized beings." [1]

If the process of development is arrested in the early stages of embryonic growth, the foetus, from the imperfection of its organization, may in many respects resemble some of the inferior animals.

The arrest of development at an early period may prevent the formation of any vestige of a particular organ ; or, if it occurs at a later period, the organ may be rudimentary.

Any severe shock of the nervous system of the mother, whether by fright or otherwise, may impair the process of nutrition, and thus produce an arrest of development in the entire embryo or some of its parts.

The rudimentary or imperfectly-developed organ may, however, attain nearly its natural size, as its growth may continue after the cessation of the developmental process.

It is likewise probable that the habitual mental condition of the mother may have an influence upon the nutrition of the embryo, and thus interfere with its development. It is well known that "a fit of passion in the nurse vitiates the quality of the milk to such a degree as to cause colic and indigestion (or

<hr>

[1] Carpenter's "Human Physiology," p. 987; "Comparative Physiology," p. 124.

even death) in the sucking infant;"[1] and it is not unreasonable to suppose that the nutritive fluids may be modified by a similiar influence during the period of gestation.

Among the abnormal conditions produced by an arrest of development are hare-lip, cleft-palate, fissures of the body or of the spinal column (*spina bifida*), absence or malformation of the limbs, deficient number of the digits, etc.

The limbs of all vertebrate animals are formed by " a kind of budding process, as offshoots of the external layer of the blastodermic membrane. They are at first mere rounded elevations, without any separation between the fingers and toes, or any distinction between the different articulations.

" Subsequently the free extremity of each limb becomes divided into the phalanges of the fingers or toes ; and afterward the articulations of the wrist and ankle, knee and elbow, shoulder and hip, appear successively from below upward."[2]

The feet of frogs, of birds, of squirrels and rabbits, of cattle, and even the feet and hands of the human fœtus, are all, in an early stage of development, webbed as if fitted for swimming, and the characteristic form of the digits in each species is only observed at a later period of growth.[3]

The divergence from this common type, observed

[1] Dr. A. Combe, on " The Management of Infants," p. 76, quoted in Carpenter's " Human Physiology," p. 1011.

[2] Dalton's " Human Physiology," p. 630.

[3] Agassiz's " Lectures on Embryology," p. 102 ; Carpenter's " Human Physiology," p. 1007 ; Colin, " Physiologie Comparée," etc., tome ii., p. 570.

in the process of development, has been made use of
in the classification of animals: as in birds, for example, those with webbed feet are placed lower in the
scale of organization, the developmental process not
having proceeded so far as in those with separate toes.

In the higher animals, digits that adhere or are
connected by a membrane represent the embryonic
type that has been retained through defective development. The lamb mentioned above, that presented
a fancied resemblance to a rabbit in the form of its
feet, was undoubtedly an instance of arrested development.

" The fœtus *in utero*, even at early periods of its
development, is liable to a large number of organic
alterations, and even to lose its life, in consequence of
inflammation attacking the uterus of the mother, the
fœtal appendages, or its own system. From such
causes arise a variety of pathological changes in the
fœtus, as atrophy, arrest of development, amputation
of limbs, and many other affections." [1]

The particular part of the fœtus affected by disease
is undoubtedly determined by the same general conditions that determine the seat of disease after birth,
among which may be enumerated irregularities of the
circulation, producing local congestions or inflammation, hereditary predisposition to disease of particular
organs, mechanical injuries, and specific diseases communicated by the mother, as small-pox, scarlet fever,
measles, etc. [2]

[1] Montgomery, "Cyclopædia of Anatomy and Physiology," article
"Fœtus," vol. ii., p. 330.
[2] Ibid., p. 333.

" According to Hausmann, the effect of variations of the external atmosphere is visible in the unusual number of blind colts and hydrocephalic pigs which are born after a wet summer." [1]

Atrophy, and even amputation of the limbs of the fœtus, has in many instances been produced by the mechanical pressure of ligamentous bands, or loops of the umbilical cord.[2]

From the facts already presented, it must be seen that malformations of the embryo are produced by well-known physiological and pathological conditions, that interfere with the normal process of development.

From what is now known of the laws of embryological development and the causes of abnormal variations, the theory that the imagination of the mother has a direct influence in producing malformations, or impressing peculiar marks upon the embryo, appears to be based on insufficient evidence.

[1] Allen Thomson, "Cyclopædia of Anatomy and Physiology," article "Generation," vol. ii., p. 475.

[2] Montgomery, loc. cit., pp. 327–330.

CHAPTER XIV.

THE causes that determine sex have been a subject of speculation from the earliest times. The theories that were first framed, in accordance with some fancied analogy, as an expression of the laws of the organization, have been repeatedly revived in their original form, without adding to our knowledge of the conditions that determine the result.

From the fact that there are two testicles and two ovaries in the higher animals, symmetrically placed, one on each side of the median line of the body, it was supposed that the right ovary and testicle were concerned in the production of males, and the left in the production of females.

Physiologists have long known that this theory had no foundation in fact, as males with one testicle and females with one ovary produce offspring of both sexes. The following case, reported by Prof. Marzolo, of Padua, is of particular interest: "In a patient, thirty-five years of age, the left ovary was removed for cystic tumor. The woman recovered from the operation, and became pregnant about a year after. She was delivered at full term of twins, a male and a female, and both of the children did well." [1]

In a case reported by Dr. Granville, of London, to the Royal Society,[2] the left Fallopian tube and ovary of a woman forty years old were entirely wanting; yet .she had been the mother of eleven children of both sexes; and, "a few days before her death, had been delivered of twins—one male and one female." [3]

" M. Jadelot, too, has given the dissection of a female who had been delivered of several children— boys and girls; and yet she had no ovary or Fallopian tube on the right side. Lepelletier asserts that he saw a similar case in the hospital at Mans, in 1825, and the *Recueils* of the *Société de Médecine* of Paris contains the history of an extra-uterine gestation, in which a male fœtus was contained in the left ovary." [4]

Mr. J. Buckingham, of Zanesville, Ohio, gives the following report of an experiment made by himself to test the truth of this theory: "Taking a boar," he says, "I took out his left testicle, and turned him into a lot with three sows, one of which had her left ovaries (ovary) out, the other the right ones (one) out, and one not spayed. The next lot had a boar with his right testicle out, and three sows fixed as the others had been. The next lot had a boar and three sows fixed as the first three had been. Now for the result: Every sow had from seven to nine pigs.

---

[1] Flint's "Physiology of Man," vol. v., p. 346; from *Gazette Médicale de Paris*, 1873, No. 44, p. 582.

[2] "Philosophical Transactions," 1808, p. 308.

[3] Dunglison's "Physiology," vol. ii., p. 400.

[4] Dunglison, *loc. cit.*, p. 410.

There were not less than three nor more than five
male pigs in every litter, or just as near half of each
as there could be." [1]

Notwithstanding these conclusive cases, experiments are now in progress of a similar character, to
test the practicability of breeding the sexes at will. [2]

A theory was advanced by Prof. Thury, of the
Academy of Geneva, that for a time was quite popular, and is now frequently advocated, although it is
readily disproved by direct observation. The first
notice of this theory in this country was published in
the *Country Gentleman*, from which we make the following quotations: "The sex depends," says M. Thury, "upon the degree of maturity of the egg at the
moment of fecundation, that which has not reached a
certain degree of maturity producing the female, and,
if fecundated when this point of maturity has passed,
producing a male."

The theory, it will be perceived, is based upon the
supposition that "the production of male organs arises
from the greater maturity and more complete development of the germ," [3] which is directly in conflict with
observations on the lower animals quoted below.

Some startling results were claimed to have been
obtained, in experiments made for the purpose of
testing the truth of the theory, but they need verifi-
cation before they can be accepted as evidence.

Although there are but few cases on record in
which exact statements are made of the facts bearing

[1] *Country Gentleman*, June, 1865, p. 364.
[2] *See Scientific Farmer*, 1876, p. 181.
[3] *Country Gentleman*, January 7, 1864, p. 12.

on this theory, the observed results of ordinary farm-
practice are sufficient to disprove it.

On the Michigan Agricultural College farm, when
under my direction, the births for ten years were as
follows: Sheep, 102.5 males to 100 females; cattle,
118.4 males to 100 females.

The system pursued for the entire time was the
same. The rams were turned with the ewes every
forenoon during the breeding-season; and the cows,
as a rule, were served as soon as they were discovered
to be in heat, the herd being frequently visited during
the day, and driven to the barn every night and morn-
ing. With very few exceptions, the females were
served during the first half of the period of heat,
which, according to the theory of M. Thury, should
have given a very large proportion of females.

The records show some remarkable facts that will
be recognized by breeders as fairly representing the
general experience of farmers. In 1864 and 1865 the
bull-calves were 2.5 to 1 heifer; in 1866 and 1867 the
heifers were considerably in excess; in 1868 and 1869
the heifers were nearly 2 to 1 bull; in 1870 the bulls
were decidedly more numerous; and in 1871 and 1872
there were more than 2 bulls to 1 heifer. In 1872
there were 2 rams to 1 ewe, and the bulls were nearly
in the same proportion to the heifers, which would
seem to indicate some peculiar influence of the season
in favor of the males. In 1871, however, the bulls
were largely in excess of the cow-calves, and there
was quite as decided a preponderance of females among
the sheep.

On Waushakum farm, according to Dr. Sturte-

vant, the Ayrshires, during nine and a half years, produced "54 bull-calves and 43 cow-calves—a proportion of 125.5 males to 100 females;" while the Ayrshires of the Oneida Community produced "26 bull
and 31 heifer calves—a proportion of 83.8 male to
100 female births." [1]

Although the last-mentioned cases have no direct
bearing upon the theory in question—the period of
heat at which copulation took place not being stated
—they are of interest in this connection, as they fairly
represent the variations that take place without apparent cause, under ordinary methods of management.

In the following cases of late impregnation, reported by Mr. Slade, the males are in excess, as they
were also in some of the instances of early impregnation above mentioned. "Three years ago last
spring," says Mr. Slade, "I had a very likely sow that
was in heat, and I let her remain thirty-six hours before taking her to the boar. The result was she had
seventeen pigs; eleven of them were males and the
others females. . . . At the next litter she remained
about the same length of time in heat before taking
the boar, and had nine male and two female pigs." [2]

On many farms the males run with the breeding
females during the season, so that copulation takes
place at the beginning of the period of heat; and yet
in such cases there is nearly an equal number of each
sex on the average, taking a number of years together,
while the males may be in excess one year and the
females in excess another year.

[1] *Scientific Farmer*, 1876, p. 166.
[2] "Massachusetts Agricultural Report," 1866–'67, p. 117.

The physiological objections to this theory, so far as any practical advantages that might be derived from it are concerned, are quite as forcible as the results of direct observation.

Fecundation, as is well known, is the result of the union of the spermatozoa of the male element with the ovum. Now, this conjunction of the male and the female elements of generation does not take place, in the higher animals at least, at the time of copulation, and it is therefore impossible to determine, in any particular case, the precise time that fecundation takes place. Observations upon the lower animals show that the spermatozoa may come in contact with the ovum in the uterus, in the course of the Fallopian tube, or at the ovary.[1]

The precise period at which the ovum escapes from the ovary is uncertain; some of the best authorities are of the opinion that the regular time for its escape is toward the termination of the period of heat, while Coste has shown that it may escape in the early part of the period, or toward its close.[2]

The conditions that determine the time of contact of the spermatozoa and the ovum are therefore exceedingly variable, and they may favor an earlier impregnation of the germ in cases of copulation toward the close of the period of heat than would be produced in other cases when copulation took place at the beginning of the period.

Experiments with dogs and rabbits show that several days may elapse after copulation before the sper-

[1] Dalton's "Human Physiology," p. 562.
[2] Ibid., p. 561.

matozoa come in contact with the ovum to produce impregnation.[1]

At a meeting of the Agricultural Society of Sé-verac, on the 3d of July, 1826, M. Charles Girou de Buzarcingues proposed "to divide a flock of sheep into two equal parts, so that a greater number of males or females, at the choice of the proprietor, should be produced from each of them. Two of the members of the society offered their flocks to become the subjects of his experiments,"[2] the results of which are given in the following table.

The principle of division was to place young rams with strong, well-fed ewes, for ewe-lambs, and a ma-tured, vigorous ram with weaker ewes, for ram-lambs.

The first experiment gave the following results:

Flock for Female Lambs served by two Rams, one fifteen Months and the other nearly two Years old.			*Flock for Male Lambs served by two Strong Rams, one four and the other five Years old.*		
Age of Mothers.	SEX OF LAMBS.		Age of Mothers.	SEX OF LAMBS.	
	Male.	Female.		Male.	Female.
Two years	14	26	Two years	7	3
Three years......	16	29	Three years.....	15	14
Four years.......	5	21	Four years......	33	14
Total	35	76	Total	55	31
Five years and over.........	18	8	Five years and over........	25	24
Total..........	53	64	Total.........	80	55
There were three twin-births in this flock.			No twin-births in this flock.		

---

[1] "Cyclopædia of Anatomy and Physiology," vol. ii., p. 465; Car-penter's "Human Physiology," p. 967.

[2] The experiments were published in the "Annales de l'Agricul-ture Française," vols. xxxvii., xxxviii., and a summary from which we quote will be found in the *Quarterly Journal of Agriculture*, vol. i., May, 1828, p. 63.

In the second experiment the ewes were divided into three sections.

The first section included the strongest ewes from four to five years old, which were better fed than the others. It was served by four ram-lambs, about six months old.

In the second section were the weakest ewes, under four or above five years old. They were served by "two strong rams," more than three years old.

The third section consisted of ewes belonging to the shepherds, "which are in general stronger and better fed than those of the master, because their owners are not always particular in preventing them from trespassing on the cultivated lands that are not inclosed." These ewes were served by the same rams as section two.

	Males.	Females.
The first section gave	15	25
" second " " . . .	26	14
" third " "	10	12

In the first section were two twin-births—four females. In the second and third there were also two—three males and one female.

These experiments were considered almost conclusive; but it will be observed that the results are not more remarkable for the range of variations presented in the relative numbers of each sex than were obtained in my experience in different years with animals under the same management.

The number of animals under observation in these experiments is too small to give the results any value as a basis of generalization, and the same objection may be made to the cases collected by Hofacker and Sadler, which we quote from Carpenter : [1]

"The following table expresses the average results obtained by M. Hofacker in Germany, and by M. Sadler in Britain, between which it will be seen that there is a manifest correspondence, although *both were drawn from a too limited series of observations*. The numbers indicate the proportion of male births to a hundred females, under the several conditions mentioned in the first column:"

	Hofacker.
Father younger than mother	90.6
" and mother of equal age . . .	90.0
" older by 1 to 6 years . . .	103.4
" " " 6 " 9 " . . .	124.7
" " " 9 " 18 "	143.7
" " " 18 and more . . .	200.0

	Sadler.
Father younger than mother	86.5
" and mother of equal age . . .	94.8
" older by 1 to 6 years . . .	103.7
" " " 6 " 11 " . . .	126.7
" " " 11 " 16 "	147.7
" " " 16 and more . . .	163.2

"From the statistics recorded in the peerages and baronetages of the United Kingdom, the proportion of male to a hundred female births is stated by Napier to be as below:" [2]

390 parents of equal age	91.8
276 fathers 1 year older than mothers . . .	101.3

[1] "Human Physiology," p. 1015.

[2] *Scientific Farmer*, 1876, p. 180, credited by Dr. Sturtevant to the *Journal of the Anthropological Society*, 1867, vol. cxix.

312 fathers	2- 3 years older than mothers	.	.	.	101.8
211 "	4- 6 " " " "	.	.	.	108.0
200 "	6-10 " " " "	.	.	.	130.1
168 "	10-16 " " " "		.	.	144.3
120 "	17-25 " " " "	.	.	.	189.7
80 "	26-32 " " " "	.	.	.	125.6
45 "	33-40 " " " "	.	.	.	112.6
18 "	40-50 " " " (mother under 25)				115.4
13 "	40-50 " " " (" over 25)				91.6

MOTHERS OLDER THAN FATHERS.

88 mothers from	1- 3 years older	.	.	.	.	94.3
77 " "	3- 5 " "	.	.	.	.	88.8
66 " "	5-10 " "	.	.	.	.	77.1
43 " "	10-15 " "	.	.	.	.	60.6
17 " "	15-22 " "	.	.	.	.	48.3

Notwithstanding the apparent uniform increase in the proportion of male births in the cases in which the father is from six to twenty-five years older than the mother, it would not be safe to attribute the variation to age alone.

In the first 1,189 cases in the table, in which the parents are of equal age, or the fathers are from one to six years older than the mothers, the average proportion of male births is below the general average, as shown by other statistics; but, as the fathers have the advantage of the mothers in age in 799 of these cases, the proportion of male births should be considerably higher. Instead of comparing the special cases collected with one another, would it not be more satisfactory to compare them with the average of all cases that can be obtained, without reference to age?

In the first 1,189 cases above noticed there would be, according to the table, 593 male children and 596

14

female children, or in the proportion of 99.5 males to 100 females.

Of the children born in Great Britain it is said that there are 104.75 males to 100 females,[1] the proportion of males being more than five per cent. larger than in the cases under consideration.

The following table, from the report of the Census Commissioners of Ireland for 1841,[2] gives the proportion of sexes in the largest number of cases in which the relative ages of the parents are stated, that has come to my knowledge :

TABLE SHOWING THE FECUNDITY OF IRISH MARRIAGES FROM 1830 TO 1841.

Description of Marriage.	AGES.		Sum of Ages of Both Parents.	Number of Marriages.	NUMBER OF CHILDREN.		Total of Children.	Proportion of Males to 100 Females.	Proportion of Children to a Marriage.
	Husbands.	Wives.			Males.	Females.			
Parties of Equal Age.	under 17	under 17	34	661	872	821	1,693	106	2.56
	17 to 25	17 to 25	42	159,761	195,895	185,913	381,808	105	2.89
	26 " 35	26 " 35	61	58,290	63,143	60,012	123,155	105	2.11
	36 " 45	36 " 45	81	3,354	1,665	1,480	3,145	112	.94
	46 " 55	46 " 55	101	428	44	51	95	86	.22
	above 55	above 55	110	136	11	6	17	183	.12
Totals and averages.........				222,630	261,630	248,283	509,913	105	2.29
Husbands who are older than their Wives.	17 to 25	under 17	38	9,847	13,203	12,558	25,761	105	2.62
	26 " 35	under 17	47	4,066	5,171	5,074	10,245	102	2.52
		17 to 25	51	128,713	159,081	150,090	309,171	106	2.40
	36 " 45	under 17	57	313	402	373	775	108	2.48
		17 to 25	61	14,325	17,478	16,270	33,748	107	2.35
		26 " 35	71	15,596	15,466	14,380	29,846	107	1.91
	46 " 55	under 17	67	36	37	37	74	100	2.05
		17 to 25	71	1,516	1,615	1,564	3,179	103	2.10
		26 " 35	81	2,469	2,109	2,011	4,120	105	1.67
		36 " 45	91	1,335	493	445	938	111	.70
	above 55	under 17	72	18	8	15	23	53	1.28
		17 to 25	76	240	198	209	407	95	1.69
		26 " 35	85	461	277	273	550	101	1.19
		36 " 45	95	429	107	79	186	135	.43
		46 " 55	105	295	14	15	29	93	.10
Totals and averages.........				179,659	215,659	203,393	419,052	106	2.33

[1] " Cyclopædia of Anatomy and Physiology," vol. ii., p. 478.
[2] Copied from Walford's " Insurance Cyclopædia," vol. iii., p. 189.

TABLE SHOWING THE FECUNDITY OF IRISH MARRIAGES FROM 1830 TO 1841.—(*Continued.*)

Description of Marriage	AGES		Sum of Ages of Both Parents.	Number of Marriages.	NUMBER OF CHILDREN.		Total of Children.	Proportion of Males to 100 Females.	Proportion of Children to a Marriage.
	Husbands.	Wives.			Males.	Females.			
Wives who are older than their Husbands.	under 17	17 to 25	38	757	996	944	1,940	105	2.56
		26 " 35	47	92	90	97	187	93	2.03
		36 " 45	57	4	......	......	.. ....	....	....
		46 " 55	67	1	......	......	.. ....	....	....
		above 55	72	1	2	......	2	....	....
	17 to 25	26 to 35	51	21,287	21,934	21,088	43,022	104	2.02
		36 " 45	61	744	372	398	770	93	1.03
		46 " 55	71	35	10	8	18	125	.51
		above 55	76	8	......	......	......	....	....
	26 " 35	36 to 45	71	2,413	1,225	1,147	2,372	107	.98
		46 " 55	81	145	40	34	74	118	.51
		above 55	85	12	......	......	......	....	....
	36 " 45	46 to 55	91	227	49	39	88	126	.89
		above 55	95	15	3	......	3	....	....
	46 " 55	" 55	105	52	2	3	5	67	.10
Totals and averages...... ...			25,788	24,723	23,758	48,481	104	1.83	
General totals and averages..			428,077	502,012	475,434	977,446	105	2.28	

When the parents are of equal age, of 509,913 children there are 105 males to 100 females.

When the father is older than the mother, of 419,052 children there are 106 males to 100 females.

When the mother is older than the father, of 48,481 children there are 104 to 100 females.

With the exception of a few instances, where the number of children is small, it will be observed that the range of variation in the proportion of the sexes, under the different conditions mentioned in the table, is exceedingly small in comparison with the variations in the preceding tables.

M. Martegoute states that at the sheepfold of the Dishley Mauchamp merinos of M. Viallet, at Blanc, the rams at the commencement of the rutting-season got more males than females; when the ewes came in heat in greater numbers, and the vigor of the ram was diminished, he got a larger proportion of females; and toward the close, when the ewes to be served were less numerous, the vigor of the ram being restored, the procreation of males was again in excess.

He concludes also that "the ewes that have produced the female lambs are, on an average, of a weight superior to those that produced the males; and they evidently lose more in weight than these last during the suckling period;" and that " the ewes that produce males weigh less, and do not lose in nursing so much as the others." [1]

I am not aware of any facts to corroborate this statement of M. Martegoute, which is not in accordance with the experience of breeders generally.

It may be that the relative age and vigor of the parents has an influence, *in connection with other conditions*, in determining sex; but that the influence is so marked as to be of any practical utility in breeding the sexes at will remains yet to be proved.

The uniformity in the proportion of the sexes, shown by statistics, in different localities, representing a great variety in the conditions of life, indicates the existence of some general law that determines the sex of offspring that is constant in its action.

The proportion of males to a hundred females in the different countries of Europe is reported as follows: [2]

[1] " Principles of Breeding," by Goodale, pp. 91, 92.
[2] "Cyclopædia of Anatomy and Physiology," vol. ii., p. 478, and " Insurance Cyclopædia," by Walford, vol. i., p. 315.

Great Britain	104.75
France	$\begin{cases} 106.55 \\ 103.38 \end{cases}$
Prussia	$\begin{cases} 106.94 \\ 105.90 \end{cases}$
Sweden	104.72
Würtemberg	105.69
Westphalia and the Rhenish provinces .	105.86
Bohemia	105.38
Netherlands	106.44
Saxony and Silesia . . .	106.05
Austria	106.10
Sicily	106.18
Brandenburg . . .	106.27
Mecklenburg	107.07
Mailand	107.61
Russia	108.91
Jews in Prussia . .	112.00
" " Breslau . . .	114.00
" " Leghorn . .	120.00
Christians " " . . .	104.00
Austria, 1830–'47 . . .	106.6
Baden, 1835–'55 . . .	105.9
Bavaria, 1835–'51 . . .	106.3
Belgium, 1841–'50 . . .	105.2
Denmark, 1835–'49 . . .	105.5
England, 1843–'52 . . .	104.7
France, 1817–'54 . . .	106.2
Hanover, 1824–'43 . . .	106.5
Holland, 1840–'53 . . .	106.5
Norway, 1801–'55 . . .	105.9
Prussia, 1816–'52 . . .	105.7
Saxony, 1834–'49 . . .	106.5
Scotland, 1855–'56 . . .	105.3
Sweden, 1749–1855 . . .	104.4

According to M. Quetelet, the proportion of sexes in Europe is 106 males to 100 females.[1] The statements in the next table were compiled by Dr. II. B. Baker, Secretary of the State Board of Health, Lansing, Michigan :[2]

LOCALITIES.	Period (Dates inclusive).	BIRTHS. Males.	BIRTHS. Females.	Total Births.	Unknown Sex.	Males to 100 Females.
Michigan..........	Year ending April 5, 1868.	10,088	9,284	19,544	172	108.6
"	Remaining 9 months 1868.	10,133	9,003	19,171	85	112.5
"	1869	14,071	12,958	27,093	64	108.5
"	1870	13,846	12,726	26,663	91	108.8
"	1871	13,596	12,327	25,992	69	110.2
"	1872	14,311	13,812	27,706	83	107.5
"	1869 to 1872	55,824	51,323	107,454	307	108.7
New York City....	1863 " 1872	121,745	113,443	235,548	360	107.3
Philadelphia.......	1820 " 1840	74,790	69,597	144,387	....	107.46
"	1861 " 1873	116,212	105,089	221,301	....	110.5
Providence........	1854 " 1872	16,145	15,351	31,496	....	105.2
Rhode Island......	1854 " 1872	42,674	40,387	83,210	149	105.6
Vermont..........	1857 " 1870	45,576	42,932	88,994	486	106.1
Massachusetts.....	1849 " 1853	73,459	68,665	142,830	706	106.9
"	1865 " 1870	107,856	101,805	209,989	323	105.9
Connecticut.......	1856 " 1871	84,436	76,797	162,510	1,227	110.0
Total and average	....	738,767	685,389	1,427,717	8,563	107.8
England..........	1850	......	......	593,422	....	104.2
"	1851	......	......	615,865	....	104.7
"	1852	......	......	624,012	....	104.6
"	1853	......	......	612,391	....	105.1
"	1854	......	......	634,405	....	104.4
Average.........	....	....	....	....	....	104.6

Variations in the proportions of the sexes from unknown causes are of frequent occurrence ; but their limited range and slight divergence from the average only show the constancy of the general law under a great variety of conditions.

" In France, during forty-four years, the male to

[1] Churchill's "Midwifery," p. 140; Carpenter's "Human Physiology," p. 1014.

[2] Michigan "Fifth Registration Report," 1871, pp. 93-112.

the female births have been as 106.2 to 100; but during this period it has occurred five times in one department and six times in another that the female births have exceeded the male." [1]

"In some districts of Norway," according to Prof. Faye, "there has been, during a decennial period, a steady deficiency of boys, while in others the opposite condition has existed." [2]

It is worthy of notice that the excess of male over female births is diminished in the case of illegitimate children.

Mr. Babbage has compiled the following table : [3]

PLACES.	LEGITIMATE BIRTHS.		ILLEGITIMATE BIRTHS.	
	Number observed.	Males to 100 Females.	Number observed.	Males to 100 Females.
France	9,656,135	106.57	673,047	104.84
Naples	1,059,055	104.45	51,309	102.67
Prussia	8,672,251	106.09	212,804	102.78
Westphalia	151,169	104.71	19,950	100.89
Montpellier	25,064	107.07	2,735	100.81
Mean		105.75		102.50

In England, according to Walford, from 1851 to 1869 the proportion of males in illegitimate births was smaller than in the case of legitimate births in the years 1851, '53, '56, '57, '60, '62, and 65, and larger in the years 1852, '54, '55, '58, '59, '61, '63, '64, '66, '67, '68, and '69, a preponderance being in favor of an increase of males in illegitimate births. [4]

[1] "Descent of Man," vol. i., p. 291.
[2] Darwin, *loc. cit.*, p. 291.
[3] Dunglison's "Physiology," vol. ii., p. 411.
[4] Walford's "Insurance Cyclopædia," vol. i., p. 315.

Of other countries the following statistics are given :

PLACES.	Legitimate Males to 100 Females.	Illegitimate Males to 100 Females.
Prussia (1820–'34).	106.0	103.1
France	106.7	104.8
Naples (1819–'24)	104.5	103 7
Austria	106.2	104.2
Würtemberg	106.0	103.5
Sweden	104.7	108.1
Bohemia	105.7	100.4
Westphalia (1809–'11)	104.7	100.4
East Prussia and Posen	105.8	103.6
Paris	103.8	103.4
Geneva (1814–'33)	109.0	101.5
Amsterdam	105.0	108.8
Leipsic	106.2	105.9
Montpellier (1772–'92)	107.1	100.8
Frankfort-on-the-Main	102.8	107.8 [1]

Amsterdam and Frankfort, with England, seem to form exceptions to the rule indicated by the statistics of other localities.

In Michigan the proportion of males to 100 females in 1870 was 108.1 for illegitimate, and 108.8 for all births; and in 1871 it was 80.0 for illegitimate, and 110.29 for all births.[2]

In Massachusetts, for twenty-three years, the proportion of male to female births, of all classes, was 107.7 to 100, while for illegitimate for the same period it was 93.4 to 100.[3]

It has been stated that "the first children of a marriage consist of a greater number of females and fewer males, in the proportion, according to Burdach, of 53 male births to 100 females,"[4] and this may per-

[1] "Insurance Cyclopædia," vol. i., p. 315.
[2] Michigan "Fifth Registration Report," 1871, pp. 93, 94.
[3] *Scientific Farmer*, 1876, p. 166.
[4] "Cyclopædia of Anatomy and Physiology," vol. ii., p. 478.

haps serve to explain the preponderance of female births among illegitimate children.

Statistics of the relative numbers of the sexes at the time of birth among domestic animals have not been published to any considerable extent, and the data that are needed for a satisfactory discussion of this subject remain in the hands of individual breeders.

The most extended collection of statistics relating to this subject within my knowledge has been made by Mr. Darwin,[1] from which the following statements are compiled:

From the limited number of cases under discussion, and the manner in which the facts have been collected, the real proportion of the sexes at birth may not be correctly represented by these statistics.

Still-born animals, and those that die at an early age, are not as a rule forwarded by breeders to the press for publication. Moreover, the records are usually made only by those who are interested in making sales of breeding-stock; and their methods of management, or the selections made for their own purposes, may have an influence in modifying the results.

Mr. Tegetmeier tabulated for Mr. Darwin from the "Racing Calendar" "the births of race-horses during a period of twenty-one years, viz., from 1846 to 1867—1849 being omitted, as no returns were that year published. The total births have been 25,560, consisting of 12,763 males and 12,797 females, or in the proportion of 99.7 males to 100 females. . . . In 1856 the male horses were as 107.1, and in 1867 as

[1] "Descent of Man," vol. i., pp. 293-300.

only 92.6 to 100 females. In the tabulated returns the proportions vary in cycles, for the males exceeded the females during six successive years, and the females exceeded the males during two periods each of four years." [1]

A writer in the London *Field* (probably Mr. Tegetmeier) in 1868 makes the statement that, " during the past four years," 3,241 fillies against 3,102 colts have been produced.[2]

Of cattle, Mr. Darwin " received returns from nine gentlemen, of 982 births, too few to be trusted. These consisted of 477 bull-calves and 505 cow-calves; i. e., in the proportion of 94.4 males to 100 females." [3]

Mr. C. N. Bement gives the record of the birth of 62 animals in 1839–'43, of which 36 were males and 26 females—a proportion of 138.4 males to 100 females.

"In another record of Short-Horn cattle, 54 bull and 52 cow calves were produced—a proportion of 103.8 male to 100 female calves." And " in another case, out of 573 entries of Short-Horn births, 235 of the calves were male, and 238 female—a proportion here of 98.3 males to 100 females." [4]

Mr. Darwin " received returns from four gentle-

---

[1] *Loc. cit.*, pp. 293, 294. In a note, p. 293, Mr. Darwin says: "During 1866, 809 male colts and 816 female colts were born, and 743 mares failed to produce offspring. During 1867, 836 males and 902 females were born, and 794 mares failed."

[2] *Country Gentleman*, September, 1868, p. 190.

[3] *Loc. cit.*, p. 295.

[4] *Scientific Farmer*, 1876, p. 166; the first item from the *American Journal of Medical Science*, October, 1845, p. 520, and the last two from the *National Live-Stock Journal*, 1872, p. 21, and 1874, p. 375.

men in England, who have bred lowland sheep, chiefly
Leicesters, during the last ten or sixteen years. They
amount altogether to 8,965 births, consisting of 4,407
males and 4,558 females, that is in the proportion of
96.7 males to 100 females. With respect to Cheviot
and Black-faced sheep," he "received returns from
six breeders, two of them on a large scale, chiefly for
the years 1867–'69, but some of the returns extending
back to 1862. The total number recorded amounts
to 50,685, consisting of 25,071 males and 25,614
females, or in the proportion of 97.9 males to 100
females. If we take the English and Scotch returns
together, the total number amounts to 59,650, consisting of 29,478 males and 30,172 females, or as 97.7 to
100." [1]

So far as numbers alone are concerned, the statistics of the sexes of sheep are more satiafactory than
of any other class of domestic animals; but there are
other facts that will undoubtedly modify the results
obtained.

In the human family, and also in many instances
with the lower animals, it has been found that the
males, at birth, are considerably in excess of the females.

The death-rate of males, however, at an early age,
is decidedly greater than it is in females, so that the
differences at birth are gradually diminished.

Maitland, in his "History of London," 1739, says:
"From the year 1657 to that of 1738, during which time
of eighty years there appear to have been christened
619,187 males and 585,334 females, and buried 994,-

[1] "Descent of Man," vol. i., p. 295.

656 males and 965,294 females, which in the christen-
ing amount to 33,853 more males than females, which
is five and a half per cent. in favor of the former; and
in the burials 29,358, which is likewise three per cent.
in favor of the males." [1]

"For every 100 still-born females, we have in several countries from 134.9 to 144.9 still-born males."
And, also, "during the first four or five years of life,
more male children die than females. For example,
in England, during the first year, 126 boys die for
every 100 girls—a proportion which in France is still
more unfavorable." [2]

In Michigan, for the year 1871, the deaths of
children under one year of age were 128.1 males to
100 females; and of those under one month, including
still-born, 138.3 males died to 100 females; and in
1870 the proportion was 121.34 to 100 females for the
first year. [3]

Carpenter states that there are on the average three
still-born males to two females; and that, of deaths
during early infancy, the proportion of males to
females is four to three during the first two months. [4]

"Several great breeders in Scotland," says Mr.
Darwin, "who annually raise some thousand sheep,
are firmly convinced that a larger proportion of males
than of females die during the first one or two

---

[1] Walford's "Insurance Cyclopædia," vol. iii., p. 204.
[2] Darwin, *loc. cit.*, p. 292, on the authority of the *British and Foreign Medico-Chirurgical Review*, April, 1867, p. 343. *See* also article "Female Life" in Walford's "Insurance Cyclopædia," vol. iii., p. 203, etc.
[3] Michigan "Fifth Registration Report," 1871, pp. 101, 170, 174.
[4] "Human Physiology," p. 1015.

years;"[1] and it is at least probable that this is true with other classes of animals.

If with sheep the enumeration of the sexes takes place at the time of castration of the rams, several months after birth, which is the common practice, the proportion of males, as pointed out by Mr. Darwin, would be too small. This element of error should not be overlooked in statistics of animals derived from published records of births, as the animals that die young, including the still-born, are not likely to be recorded; and of these the largest proportion will in all probability be males.

M. Tegetmeier tabulated for Mr. Darwin the births of a large number of greyhounds that had, during a period of twelve years, been sent to the *Field* newspaper.

" The recorded births have been 6,878, consisting of 3,605 males and 3,273 females; that is, in the proportion of 110.1 males to 100 females. The greatest fluctuations occurred in 1864, when the proportion was as 95.3 males, and in 1867, as 116.3 males to 100 females."[2]

" With respect to fowl," Mr. Darwin "received only one account, namely, that out of 1,001 chickens of a highly-bred stock of Cochins, reared during eight years by Mr. Stretch, 487 proved males and 514 females; i. e., as 94.7 to 100.

" In regard to domestic pigeons, there is good evidence that the males are produced in excess, or that their lives are longer; for these birds invariably pair,

---

[1] "Descent of Man," vol. i., p. 291.
[2] Ibid.

and single males, as Mr. Tegetmeier informs me, can always be purchased cheaper than females." [1]

As in fowls and pigeons, the sexes are not readily distinguished at an early age. Any excess in the death-rate of either sex would, as in the cases already referred to, have an influence on the results obtained.

Mr. Darwin has also collected data that render it highly probable that the males, at birth, predominate among birds, fishes, and insects. [2]

The influence of wars, famines, and epidemics, upon the birth-rate of communities is well marked in the statistics of population; but the direction in which the influence is exerted is not always the same, and in some instances is entirely unexpected. In the case of wars and of severe famines the birth-rate is diminished; while in famines of moderate severity, and in epidemics, the birth-rate is frequently increased.

When the birth-rate is diminished by war or famine for a given period, there is as a rule a decided increase in the period following that may more than compensate for the previous diminution.

Doubleday, in his "True Law of Population," says: "There are numerous instances where the occurrence of misfortune and consequent privations have given families to those who were childless in their prosperity; and, as elucidating the same law, we may adduce another fact, well known to medical persons, *which is, the extraordinary tendencies to propagation evinced by both sexes when semi-convalescent, after enfeebling and attenuating epidemics, such as fevers,*

[1] "Descent of Man," vol. i., p. 296.
[2] Ibid., pp. 296–307.

pestilences, and plagues, and the consequent extraordinary rapidity with which population recovers itself in those countries where the plague, the marsh-fever, or famines, which cause many of these epidemics, have made havoc." [1]

These influences have been considered in the chapter on "Fecundity," but they are of particular interest among the causes that determine the proportion of the sexes.

Dr. H. B. Baker has shown that the influence of the War of the Revolution and the War of 1812 are indicated in the statistics of Michigan as late as 1870, as well as the influence of the war of 1861–'65,[2] and it appears that the diminished birth-rate during these wars was accompanied by an increased proportion of male births.

It has been stated that "certain observations made by Villermé, of Paris, and by Dr. Emerson, of Philadelphia, go to show that certain causes, as great heat of summer, overworking and underfeeding, prevalence of epidemics, illegitimacy, in short, whatever tends to depress the physical and moral powers, tends also to diminish fecundity, and at the same time to reduce the excess of male births; that these causes may operate so as even to produce an excess of females." [3]

Dr. John Stockton-Hough has also made the statement that, "under ordinary circumstances, the greater

[1] Walford's "Insurance Cyclopædia," vol. iii., p. 189.

[2] "Statistics of Michigan," 1870, pp. xix.–xxi. (*See* also "United States Census," Mortality, 1860, p. 520.)

[3] "Registration Report of Kentucky," 1853, p. 119; as quoted in Michigan "Fourth Registration Report," 1870, p. 79.

the proportion of males in births the greater the fecundity." [1]

These generalizations appear to have been made without sufficient evidence, and the statistics relied upon in support of them are susceptible, in many instances, of a different interpretation.

We have already seen that depressing influences, in many cases, are favorable to fecundity; and the statistics collected by Dr. H. B. Baker seem to show that "causes tending to increase the birth-rate tend also to increase the proportion of female offspring, this being equivalent to decreasing the proportion of males." [2]

Dr. Emerson, in the paper referred to, has apparently shown that the cholera in 1832, in Philadelphia and in Paris, diminished the proportion of male births for the year 1833, and that this excess is most marked in the period of the year nine months after its "most fatal ravages." [3]

A reëxamination of the statistics used by Dr. Emerson has been made by Dr. Baker, who shows that the birth-rate was also increased in this epidemic; and he adds evidence of a similar character from the Registrar-General's report, in regard to the influence of cholera in England in 1854. [4]

The statistics of 1,427,719 births in New York City, Philadelphia, Providence, Vermont, Massachu-

[1] *Philadelphia Medical Times*, December, 1873, p. 193.

[2] Michigan "Fourth Registration Report," 1870, p. 78; Michigan "Fifth Registration Report," 1871, p. 103.

[3] *American Journal of the Medical Society*, July, 1818, pp. 78-85.

[4] Michigan "Fifth Registration Report," 1871, pp. 96-99.

setts, Rhode Island, Connecticut, and Michigan, during a series of years, as compiled by Dr. Baker, show that in 133 cases, of from 1,300 to 38,259 births according to year and locality, an increase in the birth-rate gave a decreased proportion of males, or a decrease in the birth-rate gave an increase of males; while 48 cases, on the same basis, were exceptional and 7 doubtful.[1]

Statistics of the births for 1862–'65 inclusive, in the same localities, show an increase of males, with a decreased birth-rate, resulting from the war, with the single exception of Connecticut, where there was a slight decrease in male births and an increased birth-rate.[2]

In these cases, it should be observed, the increase or decrease in the proportion of males seems to be associated with a decrease or increase of the birth-rate when compared with the average for the locality ; that is to say, the causes that apparently diminish or increase the birth-rate in a given locality seem likewise to increase or diminish the proportion of males in that locality.

It does not, however, follow from this that, in cases of remarkable fecundity, either of individuals or communities, there should be an excessive predominance of females, or that an unusual excess in the proportion of males is to be found in cases in which there is a lack of fecundity, as the conditions involved in such cases may be quite different from those that pro-

[1] Michigan "Fifth Registration Report," 1871, pp. 104–110.
[2] Ibid., 1871, p. 111.

duce variations in the birth-rate, or in the proportions of the sexes in a given locality.

The data for a satisfactory discussion of the relations of fecundity, in itself considered, to the determination of sex, are unfortunately wanting.[1]

As to the proportions of the sexes in plural births, the statistics are quite limited. In 1852, according to the registrar-general's report, "in 6,036 cases women bore two living children at a birth. In 3,587 of the above cases the children were of the

---

[1] According to Walford ("Insurance Cyclopædia," vol. iii., p. 193), the women of England, taken collectively, are more prolific than the women of Scotland; 1,000 English women, aged from fifteen to fifty-five, bearing annually 123 registered children, and 1,000 Scotch women bearing 120 children, the proportion of males to females being, according to the statistics already given, 104.7 to 100 for the former, and 105.3 to 100 for the latter.

In the vital statistics of seventeen European countries, compiled by the Belgian Government in 1866, under the supervision of M. Quetelet, it is stated that "the most remarkable rate of fecundity is shown in Russia, and especially in the single year under observation (1858), when it was nearly twice as high as reported in France" (*Statistical Journal*, vol. xxxi., p. 146; quoted in "Insurance Cyclopædia," vol. i., p. 313).

The proportion of males to 100 females in Russia for the year mentioned was 104.9, while the births in France for the same year are not given. The details of such fragmentary statistics are not sufficient to admit of any generalizations based upon them.

Statistics in Michigan show that "the birth-rate is apparently smaller among persons of African descent and larger among Indians than among the white inhabitants of this State. It also shows, what has been noticed in the two preceding "Reports," that in the case where the birth-rate is largest there is the largest proportion of female children, and where the birth-rate is smallest there is the smallest proportion of female children" (Michigan "Sixth Registration Report," 1872, p. 33).

same sex, and in the remaining 2,159 only, of different sexes."[1]

Of 457 cases of twins, collected by Churchill, both children were males in 131 cases, both females in 146 cases, and one male and one female in 179 cases.[2]

Of 56 cases of triplets, there were 3 boys in 18 cases, 3 girls in 14 cases, 2 boys and 1 girl in 11 cases, 1 boy and 2 girls in 9 cases, and in 4 cases the sex is not stated.[3]

Thirteen cases are reported of 4 children at one birth, of which there were 4 boys in 3 cases, 4 girls in 2 cases, 3 boys and 1 girl in 3 cases, 2 boys and 2 girls in 1 case, 1 boy and 3 girls in 2 cases, and in two cases the sex is not stated.[4]

In Michigan there were 389 cases of twins reported in 1870, and 298 cases in 1871, the proportion of males to 100 females being 107.14 in 1870, and 112.85 in 1871.[5]

It has generally been supposed that the production of twins was an indication of unusual fecundity; but Dr. Duncan, who is a high authority on this subject, remarks that "the variation of the frequency of twin-births in different countries is so great as to remove all probability from the notion or belief that the greater or less frequency of twins shows greater or less general fertility."[6]

[1] "Insurance Cyclopædia," vol. i., p. 318.
[2] "Theory and Practice of Midwifery," p. 402.
[3] "Insurance Cyclopædia," vol i., p. 318; vol. iii., p. 200.
[4] Ibid., vol. iii., pp. 200, 201.
[5] Michigan "Fifth Registration Report," 1871, pp. 93, 94.
[6] Duncan on "Fecundity;" quoted in "Insurance Cyclopædia," vol. iii., p. 195.

It is frequently observed that in the offspring of individuals or families there are many more of one sex than of the other, so that the determination of the sex seems to depend upon some undefined peculiarity of the parent.

Mr. Knight, from the examination of a limited number of facts of this kind, came to the conclusion that "the female parent gives the sex to the off-spring." And he says, "I have proved repeatedly that, by dividing a herd of thirty cows into three equal parts, I could calculate with confidence upon a larger majority of females from one part, of males from another, and upon nearly an equal number of males and females from the remainder." [1]

Sir John Sinclair states that "two cows produced fourteen females each in fifteen years, though the bull was changed every year. It is singular that, when they produced a bull-calf, it was in the same year." [2]

Mr. Sherman reports a predominance of heifers in the get of a Jersey bull, the cows brought to him being selected as heifer-breeders. He also states that he has "a mare that has borne ten or eleven colts, but three of which were males. They were sired by five or six stallions." [3]

Mr. Wright, Yeldersley House, says that "one of his Arab mares, though put seven times to different horses, produced seven fillies." [4]

[1] "Philosophical Transactions," 1809; quoted in Walker on "Intermarriage," p. 228.
[2] "Code of Agriculture," p. 89.
[3] *Country Gentleman*, 1877, p. 366.
[4] "Descent of Man," vol. i., p. 294.

The record of nine cows, reported by Dr. Sturtevant, "shows one set of three producing one bull and ten heifer calves; another set produced nine bull and two heifer calves; while the remaining three produced seven bull and six heifer calves." [1]

Again, the predominance of one sex may be seen in the produce of an entire herd for a single year, which may, however, be the result of peculiarities of the season or other unobserved conditions.

The Rev. W. D. Fox informed Mr. Darwin that, "in 1867, out of thirty-four calves born on a farm in Derbyshire, only one was a bull." [2]

"At the sale of Mr. Atkins's Short-Horn herd at Milcote, in 1868, it was stated, to account for the large number of bull-calves, that twenty-three of the twenty-five cows, the last season, had produced bulls." [3]

"Burdach states that those women who are most fruitful bear many more boys than girls, as in the following examples : "

		Boys.	Girls.
First woman bore . . .		26	6
Second " " in first marriage .		27	3
" " " " second "	.	14	0
Third " " . . .		38	15 [4]

These cases should, however, be classed under the head of idiosyncrasy of the parents, as there is not sufficient evidence to warrant the belief that the predominance of males is the result of extraordinary fecundity.

[1] *Scientific Farmer*, 1876, p. 193.
[2] "Descent of Man," vol. i., p. 295.
[3] *Country Gentleman*, September, 1868, p. 190.
[4] " Cyclopædia of Anatomy and Physiology," vol. ii., p. 479.

From the frequency of the cases in which there is a greater number of one sex than of the other, in the produce of particular females, it has been assumed that the female parent had a greater influence in determining sex than the male.

There are, however, several reasons why such a generalization should not be accepted as the expression of a general law. Among domestic animals, although the male practically constitutes one-half of the breeding-stock of a 'flock or herd, the females are by far the most numerous; and it would follow from this superiority of numbers that the instances of sex determined by females would be observed in greater numbers than the instances of sex determined by males, if the power of influencing sex were the same in both males and females.

Moreover, as the male is usually coupled with a number of females, any influence he might have in determining the sex of his offspring would not be so readily noticed as a similar influence on the part of the female. Again, if a particular male has a decided tendency to produce offspring of one sex, it is probable that among the many females with which he is coupled there might be found a number that have a tendency to produce offspring of the same sex.

The sex of the offspring of these females would, therefore, be determined by the combined influence of both parents acting in the same direction.

The statistics of such cases, unless great care is taken to obtain all the facts bearing upon them, might be readily interpreted as evidence that the female parent had the greatest influence upon the sex of the offspring.

If there is a great predominance of one sex in the offspring of a female by the same male for a series of years, the result may have been produced through the influence of either parent, or by the combined influence of both; but it cannot, without other evidence, be attributed to the influence of the female alone.

There are many well-authenticated cases which show that the female has not the exclusive prerogative of determining sex.

" In the ' Philosophical Transactions ' for the year 1787, mention is made of a gentleman who was the youngest of forty sons, all produced in succession from three different wives by one father, in Ireland." [1]

One of my Ayrshire cows produced one bull and five heifer calves, the bull being her first calf. Four of her daughters have produced fourteen bull-calves and one heifer—one of the daughters had seven bull-calves. These bull-calves, so far as I can trace them, differ greatly in the proportions of the sexes in their offspring, some of them getting a large proportion of females, in which they resemble their grandam, while others get a large proportion of males, in which they resemble their dams.

These cases, although not sufficient to establish any law regulating the propagation of the sexes, seem to indicate that the sex may, perhaps, have been determined by heredity, the line of descent being represented by an alternation of generations in some cases, and directly in others.

"A tomb at Ely, Cambridgeshire, England, has

---

[1] Morton's " Cyclopædia of Agriculture," vol. i., p. 337; Walker on " Intermarriage," p. 229.

the following inscription : ' Sacred to the memory of Richard Worster, who died May 11, 1856, aged seventy-three years. Also to the memory of twenty-two sons and five daughters. . . .

" The administrators of the estate of Heber C. Kimball, late Brigham Young's first counselor, filed in 1869, at Salt Lake City, a return of distributive shares, subject to the revenue tax, showing forty-one children—thirty sons and eleven daughters." [1]

A case is reported of a bull, eighteen months old, that got the first season ten heifers and two bulls. [2]

Another case is reported of a boar that " begat about seventy per cent. of males." And of another on the same farm that " got but three males out of twenty-seven pigs." [3]

Mr. Blaine says : " Some dogs, some stallions, and some bulls, are remarked for getting a greater number of males than females ; while others are the parents of more females than males." [4]

There are cases which seem to indicate that certain families may have a tendency to produce more of one sex than the other.

Sir Anthony Carlisle says, " I am intimate with a family in which the father and mother had only two children, a son and a daughter, who each married into families not related to either party, and have had fifteen daughters without one son, viz., eight by the son and seven by the daughter." [5]

[1] Walford's " Insurance Cyclopædia," vol. iii., p. 157.
[2] *Country Gentleman*, March, 1870, p. 201.
[3] *National Live-Stock Journal*, 1877, p. 101.
[4] Walker on " Intermarriage," p. 229.   [5] Ibid., p. 229.

The predominance of one sex in the offspring of particular animals has been attributed to "prepotency."[1]

This use of a term that has a definite meaning in regard to the transmission of qualities is objectionable, as it is liable to mislead those who are not aware of the special signification implied in this connection. An animal that is prepotent has a stronger influence than its mate in the transmission of its characters to their offspring. If we say that sex is influenced by prepotency, it might be inferred that the parent controlling the dominant characters of the offspring had also a predominant influence in determining sex; but this is not the case, as many instances have come under my observation in which the general characters of a pure-bred male were uniformly stamped upon his offspring out of native and grade animals; while some of the females would produce more males than females, and others would produce more females than males. The influence of individuals upon sex would, therefore, seem to depend on something that is not included in the ordinary use of the term prepotency.

There are several other theories in regard to the causes which determine sex that remain to be noticed; but the material at command will not admit of an extended discussion of their merits.

"Sir Everard Home believed the ovum or germ previous to impregnation to be of no sex, but so formed as to be equally fitted to become either male or female, and that it is the process of impregnation which marks the sex and forms the generative organs;

[1] *Scientific Farmer*, 1876, p. 193.

that before the fourth month the sex cannot be said to be confirmed, and that it will prove male or female as the tendency to the paternal or maternal type may predominate." [1]

It should be remarked in this connection that the testicles and the ovaries are formed from the same embryonic structure, and at an early stage of development it is impossible to determine which form of the generative apparatus is to be produced. According to this theory, the male offspring should resemble the father, and the female offspring should resemble the mother; but we have seen that the transmission of resemblance is frequently from the father to the daughter, or from the mother to the son; and some physiologists even claim that this is a law of heredity that has few exceptions.

Dr. Flint has presented a provisional theory that does not differ essentially from that of Sir Everard Home. He says: "It may be that when just enough of the male element unites with the ovum to secure fecundation, or when it might be said that the female element predominates, the fœtus is a female; and, when a greater number of spermatozoids unite with the vitellus, the male sex is determined.

"Such an idea, however, is purely theoretical; and the question of the determination of sex presents thus far hardly the shadow of a satisfactory explanation." [2]

Mr. Wright's directions for breeding the sexes of chickens at will is apparently based upon the idea

---

[1] "Principles of Breeding," by Goodale, p. 89; Morton's "Cyclopædia of Agriculture," vol. i., p. 336.

[2] "Physiology of Man"—"Generation," vol. v., p. 346.

suggested by Dr. Flint, that the degree of impregnation determines the sex.

Mr. Wright says: " 1. If a vigorous cockerel be mated with not more than three adult hens, the cocks almost always largely predominate in at least the early broods; later this becomes uncertain. 2. If an adult cock be mated with not more than three pullets, the result is very uncertain, the one sex being as likely to occur as the other; but usually there is a decided predominance on one side, rather than equality. 3. If an adult cock be mated with five or more pullets, the pullets are generally in excess; and what cockerels there are will be most numerous in the earlier eggs. 4. Young birds or adult birds mated together are very uncertain; but the fewer hens, and the more vigorous the stock, the greater is the proportion of cockerels, which are always more numerous in the earlier eggs of a season than the later." [1]

As Mr. Wright, however, admits that "there will be numerous and startling exceptions" [2] to these rules, they can hardly be accepted as the expression of a general law.

The theory presented by Dr. Flint appears to be in direct conflict with apparently well-authenticated facts observed among insects. In a hive of bees may be found a queen-bee (a perfect female), a number of drones (males), and the neuter workers (imperfect females), which are by far the most numerous. It is well known, as first shown by Dzierzon, that an unimpregnated queen lays eggs that produce drones, and

---

[1] "The Illustrated Book of Poultry," p. 133.

[2] *Loc. cit.*, p. 45.

that the workers, although incapable of impregnation, may sometimes lay eggs that produce drones. Eggs producing females are, however, only laid by the queen, and then only after impregnation.[1] It appears that all unimpregnated eggs produce males, and all impregnated eggs produce females—i. e., either workers or queens: and the male element of fertilization, in the case of bees, would seem, therefore, to be essential to the production of females only.

The theory that sex is determined by the activity of the nutritive processes has been recently advocated by naturalists.

Mrs. Mary Treat, in the *American Naturalist*, has given the results of a large number of experiments with butterflies, showing that, if the larvæ are not well fed before going into the chrysalis state, the perfect insects developed from them are males ; but, if the larvæ are abundantly supplied with food, the perfect insects are females.[2]

In a paper communicated to the Philadelphia Academy, Mr. Gentry details a series of experiments with the larvæ of various species of moths, the results of which agree with those obtained by Mrs. Treat with butterflies. Mr. Gentry arrives at the following conclusions: " 1. That males are the invariable result when the larvæ are fed on diseased or innutritious food ; 2. That in the fall, when the leaves have not their usual amount of sap, males are generally produced ; 3. That more males are produced late in the

---

[1] " The Dzierzon Theory," by Baron Berlepsch, pp. 13–36 ; " Hive and Honey-Bee," by Langstroth, pp. 40–45 ; and other works on the bee.

[2] *Popular Science Monthly*, June, 1873, p. 252.

season than females; 4. That the sexes in early life cannot be distinguished, the change being brought about late in life by the conditions of nutrition." [1]

Mr. Thomas Meehan has made observations which seem to show that "sex in plants is the result of the grade of nutrition, the highest grades of nutrition or vitality producing the female sex, and the lower grades the male." [2]

These changes in the reproductive organs that are produced by conditions of nutrition appear to be analogous to those that determine the development of the reproductive organs of the queen-bee.

If a queen is destroyed or removed from the hive, "the bees choose two or three from among the neuter eggs (producing workers) that have been deposited in their appropriate cells, and change these cells (by breaking down others around them) into *royal* cells, differing considerably in form, and of much larger dimensions; and the larvæ, when they come forth, are supplied with '*royal jelly*,' an aliment of a very different nature from the ' bee-bread ' which is stored up for the nourishment of the workers, being of a pungent, stimulating character. After going through its transformations, the grub thus treated comes forth a perfect queen, differing from the ' neuter,' into which it would otherwise have changed, not only in the development of the generative apparatus, but also in the form of the body, the proportionate length of the wings, the shape of the tongue, jaws, and sting, the absence of the hollows on the thighs in which the

---

[1] *Popular Science Monthly*, April, 1874, p. 762.
[2] Ibid., p. 761.

pollen is carried, and the loss of the power to secrete wax. Thus, in acquiring the attributes peculiar to the perfect reproductive female, the insect loses those which distinguish the working population of the hive; and, of this departure from its usual mode of development, the difference in the food with which it is supplied appears to be the only essential condition." [1]

In the development of a queen from a worker larva, it does not appear to be necessary that the changed conditions, as to form of cell and nutriment, should be made at the earliest period of growth. Dzierzon says: "I have noticed that worker larvæ, so far advanced that they nearly fill their cells, will still be developed as perfect queens, if, before capping, the cell be somewhat enlarged and widened, and the larvæ supplied with the appropriate pabulum." Baron Berlepsch adds, "Incredible as this at first seemed, I have found it, nevertheless, true." [2]

It is possible that the influence of heat and light upon the sex of plants, observed by Mr. Knight, may be owing to changes produced in the nutrition of the plant.[3]

Mr. Knight says: "I can at any time succeed in causing several kinds of monœcious plants to produce solely male or solely female blossoms. If heat be, comparatively with the quantity of light which the plant receives, excessive, male flowers only appear; but, if light be in excess, female flowers alone will be produced." [4]

---

[1] Carpenter's "Comparative Physiology," p. 163.

[2] "The Dzierzon Theory," by Baron Berlepsch, p. 45.

[3] "Comparative Physiology," p. 618.

[4] "Physiological and Horticultural Papers," p. 358; quoted in *Scientific Farmer*, 1876, p. 181.

The relations of the function of nutrition to that of reproduction may perhaps explain some of the cases that have been cited as evidence in favor of other theories; and it may be that the determination of sex depends upon a number of conditions that are all intimately connected with the function of nutrition.

Another theory has been recently presented by Mr. John R. Stuyvesant, of Poughkeepsie, New York, that is evidently based upon two assumptions, viz.: first, that the sex of the offspring depends entirely upon the female; and, second, that every alternate egg is of the same sex. The cases given above are sufficient to show that these assumptions are without foundation, while the limited number of cases that have been presented as evidence of the value of the system cannot be accepted as conclusive. Mr. Stuyvesant says: " My plan is simply this : if a cow has produced for her last calf a heifer, I do not allow her to be served the first time she comes in season, but let her run over until the second time, when she is served in the first part of her heat, and is immediately shut up by herself until it passes over. Should she not catch this time, I let her run over heat number three and serve her in heat number four, and so on until she finally does catch.

" If a cow has last produced a bull-calf, then, in this case, I have her served the very first time she comes around after calving, and shut her up by herself as in the preceding case. Should she not catch by this service, I let her run over the next, or season number two, and serve her the next, or season number three, etc., until she catches. My reasons for so doing are just these : I take for granted that every alternate egg or ovum presented for impregnation is a male. Consequently, if a cow has a bull-calf, the next egg in her rotation must be a female, and, if impregnated when presented, the produce will be a female, etc." [1]

It will be difficult to reconcile the hypothesis presented by Mr. Stuyvesant with the cases in which the sex of the offspring is apparently determined by the male parent, the proportions of the sexes observed in plural births, or with the observed influence of nutrition upon sex in insects, recorded by Mrs. Treat and Mr. Gentry.

In the last-mentioned cases the sex was not determined until the embryo had reached an advanced stage of development, and the eggs could not, therefore, have been endowed with sexuality ; and we have no physiological reasons for the belief that the ovum of the higher animals presents an exception, in this respect, to the general law that governs the function of reproduction.

[1] *Country Gentleman*, 1877, p. 415.

CHAPTER XV.

A PEDIGREE is a record or statement of the ancestors of an animal, that serves as a guide in tracing inherited characters.

In itself considered it is not necessarily an evidence of purity of blood, as animals of mixed blood may have a recorded pedigree as well as those that are purely bred.

The first records of animals belonging to the different breeds are to some extent conventional, and the details of the lineage are not always given with the accuracy that is required in recording their descendants.

When animals in a particular locality have certain general characteristics which they transmit with uniformity to their offspring, they are recognized as a distinct breed. The descendants of these animals may be bred for an indefinite period without any published record of their ancestry, until the importance of the breed and its wider diffusion render it desirable that a systematic record be made, that will define its limits and enable breeders to readily trace the various lines of descent that connect their animals with the original representatives of the breed. When sufficient encouragement is given to insure the success of the enterprise, an individual, or an association of breeders, may undertake the publication of a record, which is called a herd-book or stud-book.

Any animals that are generally acknowledged to belong to the particular breed are admitted to this record, although their owners may not be able to furnish a detailed statement of their ancestral history.

Animals that have been the means of establishing the reputation of the breed by their superior merit, will be found on the record, side by side, not only with the inferior members of the breed, but with those of questionable purity of blood. Many animals may trace their descent from herds that have been noted for producing the best representatives of the breed, while others will have nothing in their ancestral history to recommend them aside from their supposed purity of blood.

As the original records include animals of very unequal merit, their descendants, that appear in the later volumes of the herd-books, must present a like diversity in their qualities.

From these differences in the quality of the animals entitled to record, it will be seen that the inherited peculiarities of an individual, aside from the general characters belonging to the breed, must be determined by evidence not contained in the herd-books; and that the pedigree, as recorded, will only serve as a guide to the study of inherited characters, from its enumeration of the individuals comprised in the ancestry, without indicating their relative rank or value.

The value of any pedigree will depend upon its authenticity, completeness, and the quality or characteristics of the animals comprised in the ancestry.

The authenticity of a pedigree is to be determined by the same rules that guide us in deciding upon the truth of any other statement or record. In the first place, it must be consistent with itself, and in accordance with the known facts in the history of the breed.

The reputation of the breeder for integrity, and the care with which he keeps a record of the breeding of his stock, together with his opportunities for obtaining correct information in regard to the statements he places on record, must all be taken into consideration. A pedigree made from memory alone, some time after the occurrence of the facts recorded, cannot be so satisfactory as one based on records made at the time.

The intentional or careless omission of a name in copying a pedigree is a frequent source of error that is not easily detected by persons who are not familiar with the history of the family to which the animal belongs.

Under the head of completeness it is important that the name and residence of the breeder and the present owner be given, together with the date of birth, the color, and other distinguishing marks that may aid in identifying the animals that are named in the record.

Every animal mentioned in the pedigree should be traced through every line of descent to individuals of acknowledged purity of blood.

A defect in the record, or the evidence of a cross

of impure blood several generations back, may appear to be of but little consequence from the very small fraction that apparently represents the proportion of impure blood in the system.

The facts of atavism, and the observed influence of a cross for many generations afterward, will, however, show that the intensity of an inherited peculiarity cannot be expressed or represented in mathematical terms.

Fleischmann states[1] that the common sheep in Germany grow from " 5,000 to 5,500 wool-hairs " to the square inch; while the pure-bred merino sheep that are used in improving them by crossing have from " 40,000 to 48,000 wool-hairs " to the square inch. The cross-bred sheep, when a pure merino ram has invariably been used on one side of the ancestral line, have but " 27,000 wool-hairs " to the square inch " in the twentieth generation," which is about a mean of the numbers observed in the common sheep and the merinos.

If the " blood " of the original varieties had been transmitted in mathematical proportions, a grade or cross-bred of the twentieth generation would have less than one-millionth part of the " blood " of the common sheep. The number of wool-hairs to the square inch, and other peculiarities of the wool in such cross-bred animals, show that this apparently insignificant fraction of blood has a marked influence on the character of the fleece.

The completeness of the ancestral record, and the unquestionable purity of blood of every animal in-

[1] " Patent-Office Report," 1847, pp. 269-271.

cluded in it, would therefore seem to be a matter of real importance.

As there are many animals of the same name recorded in the herd-books, the recorded number, which becomes a part of the name itself, must be given in the pedigree as the only means of identification.

As animals may be descended in one or more lines from a given herd, and still have an infusion of blood from other sources that may be objectionable, a pedigree should not end in a general clause indicating descent from a specified herd or importation, without giving in full the name and number of each animal in every line of descent.

A pedigree tracing all lines of descent from animals bred by men who were known as breeders of pure-bred animals at the time the first records were made must, however, be accepted as complete, as all recorded pedigrees have a similar basis.

After examining a pedigree with reference to its authenticity and completeness, the characteristics of the individuals included in each line of descent, and of the families which they represent, should be carefully considered.

If all the ancestors of an animal have been remarkable for their good qualities, and their conformity to the same general type of excellence, and for their freedom from serious defects, its inherited peculiarities will be valued not only for the merits it may be expected to possess as an individual, but for the certainty with which the dominant characters of the family will be transmitted to its offspring. If, on the other hand, the ancestors present great variations of form and quality, with the frequent occurrence of defects that diminish the value of an individual for a special purpose, the dominant characters of the animal, as determined by inheritance, cannot be predicted with any certainty, and it cannot be relied on to transmit the most desirable qualities of its ancestors, as dominant characters, to its offspring.

The great difference in the actual value of animals, arising from their inherited qualities, may be seen from a single illustration :

Two persons, A and B, begin to breed Short-Horns at the same time, by making a selection of females from the same herd, so that the value and quality of the animals they start with are the same.

Mr. A, who has definite ideas of the form and qualities of the animals he proposes to breed, makes a careful study of his herd, with reference to the selection of a bull that will correct any defects he may observe, and, at the same time, improve it in one or more of the most desirable characters. After visiting a number of herds, and making a considerable expenditure of time and money, he finds the bull that will best suit his purpose; but, as it is an animal of extraordinary merit, the owner does not propose to part with it. As this is the only animal Mr. A has seen that answers all the requirements of his standard of excellence, he finally secures his services at what others might consider an extravagant price. He is determined to establish a herd of the highest attainable excellence, and does not count the cost in carrying his designs into effect.

After using this bull in the herd as long as it

seems desirable, another, of the same general type and qualities, is selected with the same care and judgment, and regardless of expense, the only consideration that determines his choice being the usefulness of the animal in the improvement of the herd. When better animals of the desired type cannot be obtained elsewhere, some favorite male of his own herd is selected to perpetuate the valuable qualities already established.

After practising this system of selection rigorously for many years, the herd may become celebrated for its uniformity and excellence, and other breeders will find it for their interest to resort to it for males to improve their own stock, that has not been so carefully bred.

Mr. B, with the same opportunities for improvement, has no definite standard of excellence, and is, moreover, unable to detect the defects of the females he has selected as the foundation of a herd. In the selection of a male he might consider purity of blood the most important, and perhaps the only, consideration. As he does not realize the fact that the male, in effect, constitutes one-half of the breeding elements of his herd, he may, perhaps, think he cannot afford to buy a high-priced bull, the purchase already made having nearly exhausted his supply of ready money. A neighbor has a pure-bred bull he does not wish to use any longer, which he offers to sell at one-half the price originally paid for him.

This bull may have all the defects and but few of the good qualities of the females already purchased, yet he is placed at the head of the herd simply because he can be bought at a moderate price. Some of his calves may resemble their dams in their best points, while in many of them the defects of both parents may predominate.

When another bull is needed in the herd, a similar selection is made, in direct violation of the established rules of the modern system of breeding.

The effects of this hap-hazard system, or, rather, lack of system, are readily recognized in the great differences in form and quality presented by individual animals, and the low average excellence of the herd.

The pedigree of an animal from the herd of Mr. A would not only represent qualities that were in themselves valuable, but a potency in the hereditary transmission of these qualities that would be highly valued by the experienced breeder.

The pedigree of an animal from the herd of Mr. B would not add to its value for the purposes of the breeder, as its inherited tendencies, as shown by its ancestral history, would be such as it would not be desirable to perpetuate.

From these extreme cases it must be seen that the value of an animal for breeding purposes does not depend entirely upon its form and apparent qualities when studied as an individual, but also upon its pedigree which represents the sum of its inherited characteristics.

In breeding-stock, individual excellence in connection with the best inherited characters is of course desirable in all cases ; but, when it is impossible to secure this combination of qualities, the breeder should not lose sight of the fact that the greatest perfection in the individual will not compensate for ancestral defects that have been frequently repeated, as the latter will in all probability have a predominant influence upon the offspring.

In this connection, it may be well to notice what are popularly called "fancy prices" and "fancy points" in breeding. Many persons who are not familiar with the practice of the best breeders seem to think that almost everything relating to pure-bred stock is a matter of fancy only, and that the qualities of real value for the practical purposes of the farm are neglected.

It is undoubtedly true that breeders often fail to comprehend the relations of form to the qualities of intrinsic value, and that errors in judgment in the selection of breeding-stock are of frequent occurrence.

The breeders who have gained a world-wide reputation in the improvement of the different breeds have, however, made the development of useful qualities their leading or sole object. The animals that have been sold or "let" at extraordinary prices have been members of families that were noted for their uniform good qualities, and for their prepotency in transmitting their characteristics when coupled with animals that represent the average excellence of the breed. Their real merits as breeders made them "popular," or "fashionable," and the high prices that they have commanded have been largely the result of competition among breeders who were seeking the best means of improvement in their flocks and herds.

It must, however, be admitted that in many instances the prices of valuable animals have been enhanced by mere speculators, who knew little and cared less for the true principles of breeding, and the improvement of animals for their useful qualities.

That the real interests of those who are engaged in the legitimate business of breeding have been injured by the speculative buyers of favorite families, no one familiar with the history of our improved breeds will deny.

Even breeders of ability, who might gain an enviable reputation, and a satisfactory pecuniary reward, by devoting their energies to the development of the best qualities of their favorite breed, have encouraged the prevailing mania for speculation by making purchases of animals that could not possibly be of use in the improvement of their herds, and putting them up at auction with the best of their own breeding to make an attractive sale.

Notwithstanding the extravagant prices that have been paid, under the stimulus of excitement, for animals that were of but little value for any practical purpose, persons who have a full knowledge of the interests involved in the legitimate business of breeding choice stock would hardly be willing to set a limit to the prices that may be consistently paid or refused for animals of extraordinary merit, that are especially adapted to the wants of a carefully-bred flock or herd.

The late Edwin Hammond remarked to the writer, after refusing what appeared to be an extravagant price for his ram Gold Drop, that he could not afford to sell his best ram at any price, unless he should decide to give up the business of sheep-breeding, and sell his entire flock.

With the true spirit of a successful breeder, who had spent many years of systematic effort in establishing a flock that excelled in its useful qualities, Mr. Hammond placed a higher value upon the improvements he expected to obtain in his entire flock by the use of this ram than he did upon the money he could have been sold for.

The breeder who can be tempted by high prices to part with his best animals cannot reasonably expect to succeed in establishing a flock or herd of remarkable excellence.

Peculiarities of color or form, that do not represent any valuable qualities, may be properly called "fancy points;" and the money paid for them by the purchaser may be considered a "fancy price," whether the amount is small or large.

The solid color and black points of the Jerseys, and the red and dark roan of the Short-Horns, that are so fashionable at the present time in America, are good illustrations of mere "fancy points," that should not be taken into consideration in forming a consistent standard of excellence.

The prevailing fashion for particular colors, without reference to the qualities connected with them, must not only tend to retard the improvement of these breeds, but to diminish their value for practical purposes by encouraging the selection of breeding-stock in accordance with a false standard of excellence.

The wrinkles on merino sheep furnish another illustration of fancy points that are not only useless in themselves, but decidedly injurious, from the blending of different styles of wool in the fleece that

diminishes its value for the purposes of the manufacturer.

The interests of the breeder of fine-wooled sheep
will always be best promoted by the production of a
style of wool that is adapted to some special purpose,
and that will, therefore, command the highest price
in the market.

A peculiarity that is characteristic of the breed, or
of a family of extraordinary excellence, although triv-
ial in itself, would, however, be of real value as an
indication of the inheritance of the qualities of the
breed or family, and could not, therefore, be considered a fancy point.

The Southdowns bred by Mr. Webb, which were
justly celebrated for their superior qualities, had lighter-
colored faces and legs than other families of the breed,
and the breeders at the present time value the lighter
shades of color of the face and legs as representing
one of the characteristics of the Webb blood.

The tan-colored marks on the ears and faces of
the merino sheep were highly prized by the early
breeders as an indication of the "blood," and they
could not, strictly speaking, be called "fancy points,"
as they represented inherited tendencies that were in
themselves valuable.

For convenience and exactness of expression in
discussing the lineage of animals, and in recording
pedigrees, the following terms are in use among
breeders :

The term "thorough-bred," in its strict signification, is used to designate the English race-horse, and it
has been generally adopted as the name of the breed.

In America the term "thorough-bred" is frequently applied to sheep and cattle; but, as there are other terms in use to express the same idea that are quite as definite and concise, without being open to the objection of ambiguity, it would be well to restrict it to its original use as the name of a distinct breed of horses.

"Pure-bred," "full-blood," and "thorough-bred," were defined by the American Association of Short-Horn Breeders as synonymous terms, and to indicate "animals of a distinct and well-defined breed, without any admixture of other blood." [1]

The following definitions were also adopted by the association:

"'Cross-bred'—animals produced by breeding together distinct breeds.

"'Grades'—the produce of a cross between a 'pure-bred' and a 'native.'

"'High-grade'—an animal of mixed blood, in which the blood of a pure breed largely predominates."

Close-breeding is the coupling of animals that are closely related; while "in-and-in breeding" implies the closest possible relationship in the animals bred together.

High-breeding is sometimes used as synonymous with close-breeding, but it properly signifies a rigorous selection of breeding-stock with reference to a definite standard, and within the limits of a particular family.

In what is popularly called "breeding-in-the-line,"

[1] "Proceedings of American Association of Breeders of Short-Horns," Indianapolis, 1872, p. 21.

the selection of males is limited to a particular family, without reference to the quality or uniformity of the animals selected. Strictly speaking, however, it means the selection of males of a common type and belonging to the same family.

In defining the parentage of animals, the terms "out of" and "got by," or, in the abbreviated form, "by," are made use of, the former referring to the dam and the latter to the sire; for example, Favorite (252) was got by Bolingbroke (86), out of Phœnix by Foljambe (263): that is to say, Bolingbroke (86) was the sire of Favorite (252), and Phœnix by Foljambe (263) was the dam of Favorite (252); and the sire of Phœnix was Foljambe (263). The term "out of" is sometimes improperly used in referring to the sire; it should, however, for the purpose of exactness, be used only when referring to the dam; and "got by" or "by" should be as strictly limited to a reference to the sire.

In Short-Horn and Hereford pedigrees the bulls only have a number, while the females are designated by the name of their sire following the word "by;" as, in the above example, the cow Phœnix is distinguished from all others of the same name by being the daughter of the bull Foljambe (86).

As there are several Short-Horn herd-books, it becomes necessary to indicate in the pedigree the particular record to which the numbers attached to the names of bulls refer.

Where some other method is not specified, numbers without distinguishing marks are understood as referring to "The American Herd-Book;" those placed within marks of parenthesis ( ) refer to " The English Herd-Book," as in the above examples; while the numbers of " The Canadian Herd-Book " are placed within brackets [ ].

When numbers refer to Alexander's " Short-Horn Record," the method of distinguishing them is usually mentioned in connection with the pedigree.

In the Devon, Jersey, and Ayrshire herd-books, the females, as well as the males, have a distinguishing number.

For the convenience of those who are not familiar with recorded pedigrees, examples showing the form of record in the different herd-books will be found in the Appendix.

CHAPTER XVI.

THE constitutional tendencies and general characteristics of animals may be ascertained, as we have seen, with great certainty by a study of their ancestral history. Any additional information in regard to the details of the organization, which determine the qualities that are of value in the economy of the farm, as the disposition, nervous energy, muscular strength and activity, quality of flesh, proportion of valuable carcass, activity of the processes of nutrition, and strength of constitution, must be gained through the indications presented in the external form, that are manifest to the sight and touch.

Too little attention has been paid to the relations existing between the external form of the animal and its internal and more obscure characteristics, upon which its value in a great measure depends.

Every part of the external conformation should be associated in the mind of the breeder with the correlated peculiarities of structure that give the greatest value to the animal for some particular purpose, and thus serve as an index to the many important characteristics that might otherwise escape attention.

The eye should be trained to detect the slight

modifications of form that indicate real values, and our notions of beauty in external form and expression should be based upon an assemblage of symmetrical characters that are in themselves useful.

Without some consistent standards of beauty and utility, that have a definite relation to the details of the organization, individuals will unavoidably differ in opinion, not only as to what constitutes perfection of form, but as to the relative value of the different parts of the body, which, when taken together, give expression to the general conformation of the animal.

Any general expression of symmetry or proportion that is pleasing to the eye may be regarded, by the unskilled observer, as a form of beauty that is satisfactory, although it may not represent any of the qualities that render the animal valuable for any useful purpose.

The practical man, looking upon pecuniary values as the true standard of excellence, will only be pleased with the symmetrical proportions of form that indicate the presence of valuable qualities in the greatest perfection.

In the improved breeds the peculiarities of form and character that adapt the animal to a particular purpose are most highly prized, and the relative value of individuals therefore depends, to a great extent, upon their development in a special direction.

The principle of correlation that enables the breeder to determine the internal characteristics and tendencies of the organization, through the indications presented by the external form, is of general application, and may be made use of in the study of animals

16

representing the different breeds; but the points upon
which an opinion is formed will necessarily have a
different value in each breed, from the difference in
the qualities that constitute perfection.

In animals intended for the butcher, the most
satisfactory test of merit can only be applied when
they reach their destination on the *block*, where the
relative development of the most valuable parts can
be readily demonstrated.

This test, from its very nature, cannot, however,
be applied in those cases in which a reliable method
of estimating real values is most needed—as in determining the relative merits of breeding-stock, or the
feeding qualities of animals that are to be fattened.

As a practical test of the true value cannot, in
many instances, be applied to the living animal—as
in determining the greatest proportion of choice parts
in animals intended for the butcher—the prospective
value of young animals for the dairy or for work—or
the ability of animals in the lean condition to fatten
rapidly when well fed—we must resort to the ancestral
history for a knowledge of inherited tendencies, and
to the details of external conformation for an index
of all other particulars.

To become an expert in judging animals with
reference to their value on the whole, for a particular
purpose, requires extended opportunities for observation under a variety of conditions—a careful study of
their form when alive, in connection with their appearance on the butcher's block, where the leading
object is meat, and for other purposes, the relations of
their form to the activity of the functions concerned

—in the performance of labor, or in the production of wool and milk—must not be overlooked, and even then a long experience will be required to train the eye and the touch to make nice discriminations in essential details.

Admitting, then, that the thorough knowledge of animals that enables a person to form a correct opinion as to characteristics and quality cannot be gained without practical training and experience, it is nevertheless true that the acquisition of such knowledge may be facilitated by a study of the correlated structure of the animal organization, so that the relative value of different parts, and the relations of one organ or set of organs to another, and to the entire system, may be clearly understood.

The principles that are applied in the study of comparative anatomy and physiology, in tracing the harmonics of structure and function in allied groups, which have been discussed in a preceding chapter, must then be of practical interest to the breeder, as they aid him in determining the relative value of the various modifications of form observed in the animals he is trying to improve.

Moreover, the external form and proportions of an animal, when studied from this point of view, cannot fail to furnish the most satisfactory indications of the structure and functional activity of the internal organs concerned in the complex processes of nutrition, upon which all forms of animal products depend.

As the greatest excellence in the production of meat, or milk, or wool, or labor, involves peculiarities of structure and function that adapt the animal in each case to a special purpose, it will be necessary to consider separately the correlations existing in these different forms of production.

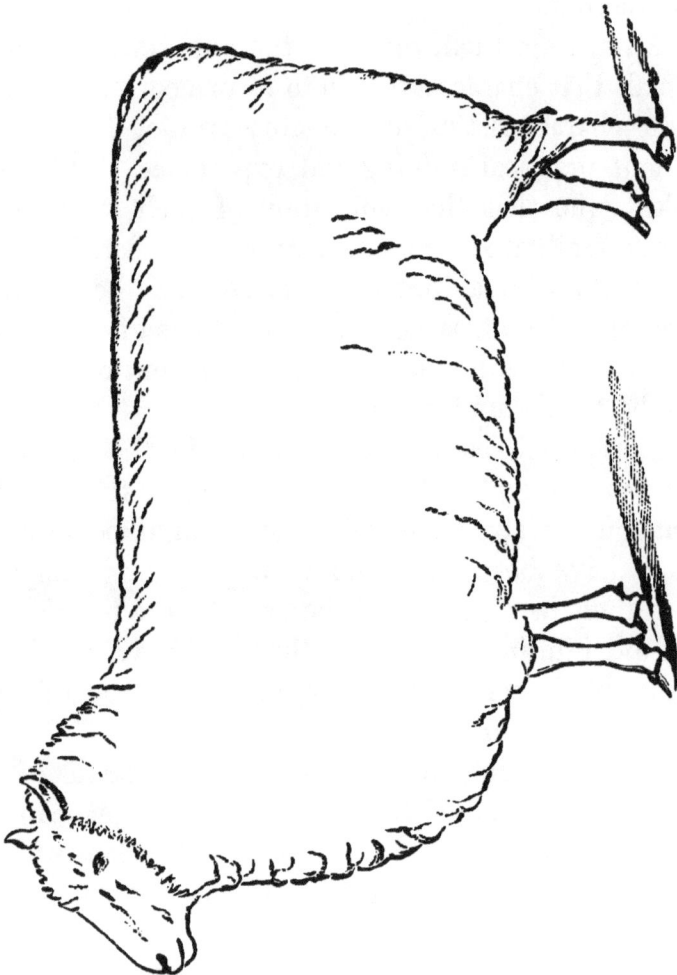

FIG. 1.—SOUTHDOWN.

All animals belonging to the best developed meat-producing breeds have essentially the same general characters and form, and a corresponding similarity prevails in their correlated structure.

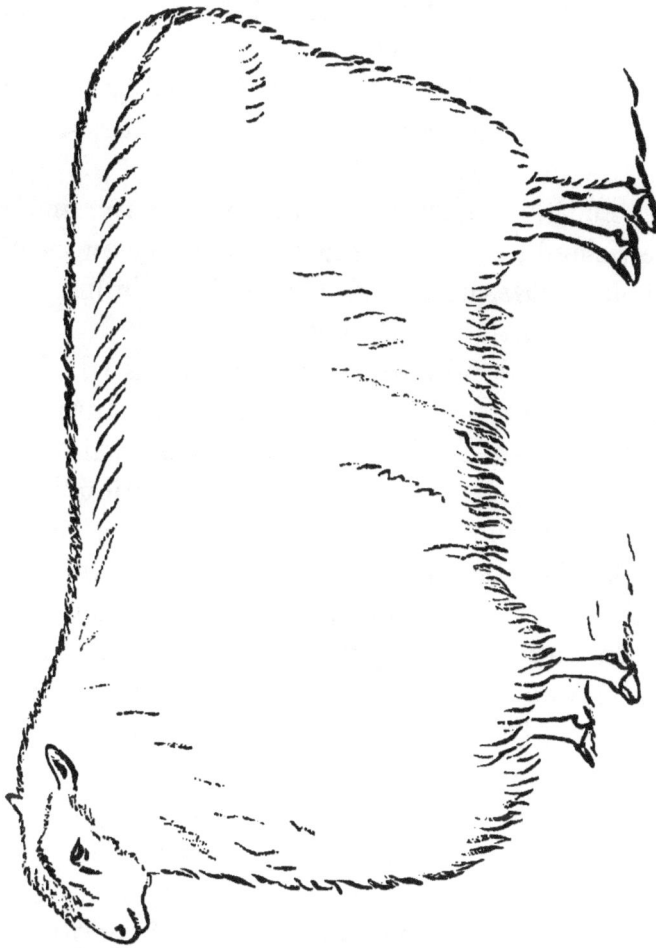

FIG. 2.—COTSWOLD.

The following characteristics may be mentioned as of especial importance, the absence of any one of them tending to materially diminish the value of the animal in the production of meat :

1. A sound constitution is of course desirable in all animals, but it is indispensable in the feeding animal whose powers of nutrition are taxed to the fullest extent in the rapid conversion of the food that is required in successful feeding.

2. Good feeding quality, or the ability to fatten rapidly at an early age and return the largest profit for food consumed.

3. The flesh should be of good quality, and the carcass should furnish the largest possible proportion of choice parts, with a corresponding diminution of the parts of little or no value.

The general proportions of the animal which first naturally attract attention will frequently furnish indications of its leading characteristics, without an examination of the details of its conformation.

Many of the best authorities on external form agree in the statement that the body of an animal intended for the butcher should be somewhat rectangular in outline, giving the form of a parallelogram when viewed from the side, and of a square when viewed from before or behind ;[1] but, in approximating to these mathematical figures in outline, it should be remembered that the angular parts of the body must be rounded and smoothly blended with the general surface, without any bony prominences or coarseness to detract from the general expression of compactness, substance, and symmetry, that marks the perfection of useful beauty.

[1] *Quarterly Journal of Agriculture*, vol. v., p. 162, vol. vi., p. 267 ; Johnson's " Farmer's Encyclopædia," p. 299 ; *Farmer's Magazine*, vol. xxxix., p. 480, vol. ii., p. 97.

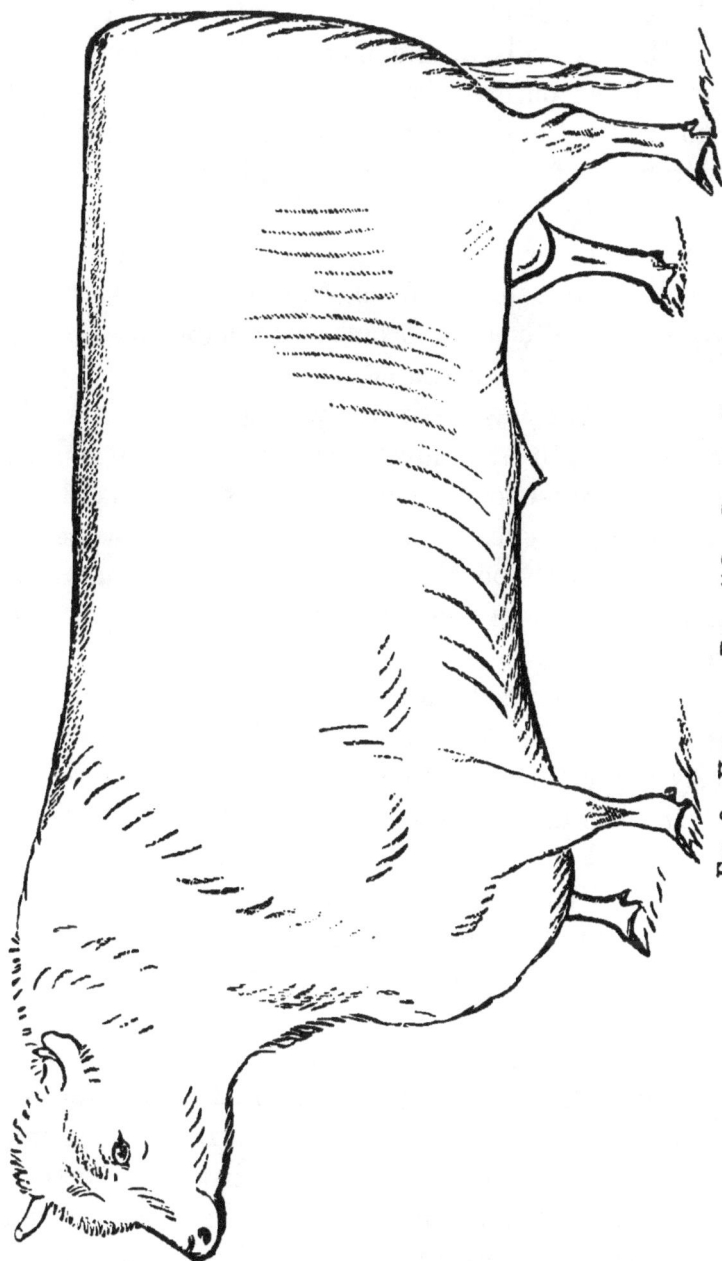

FIG. 3.—HEREFORD BULL "SIR CHARLES."

Several illustrations of the rectangular type of form that prevails in the meat-producing breeds are given in the outline sketches, Figs. 1, 2, 3, 5, 6, and 7; in contrast with them, Figs. 8 and 9 furnish good illustrations of forms that are objectionable.

Regularity and symmetry in the general outline are not, however, sufficient in themselves to constitute perfection in external form, the proportions of the body being quite as important as an indication of the characteristics of the animal.

If the body is excessively long, without corresponding depth and substance, and the under-line, from the proportionate length of legs, is too far from the ground, a delicacy of constitution is indicated, in connection with poor feeding quality, late maturity, and a deficiency in the proportion of choice parts in the carcass.

The long-bodied bull, Fig. 4, sketched from life, has nearly all the defects that usually accompany such faulty proportions in general form. The chest is narrow and lacking in capacity, as indicated by the form of the brisket, the defective fore-flank, flat ribs, and deficient girth; the shoulder is too upright, the crops defective, the loins narrow, the flanks light, and there is too large a proportion of the coarser parts of the carcass.

A low and remarkably short body, with great depth and thickness of carcass, as in Fig. 5, indicates a tendency to mature early, to lay on fat rapidly, and it may be in excess and in masses that are objectionable, with a deficiency in muscle or lean meat. From their extreme compactness such animals may weigh

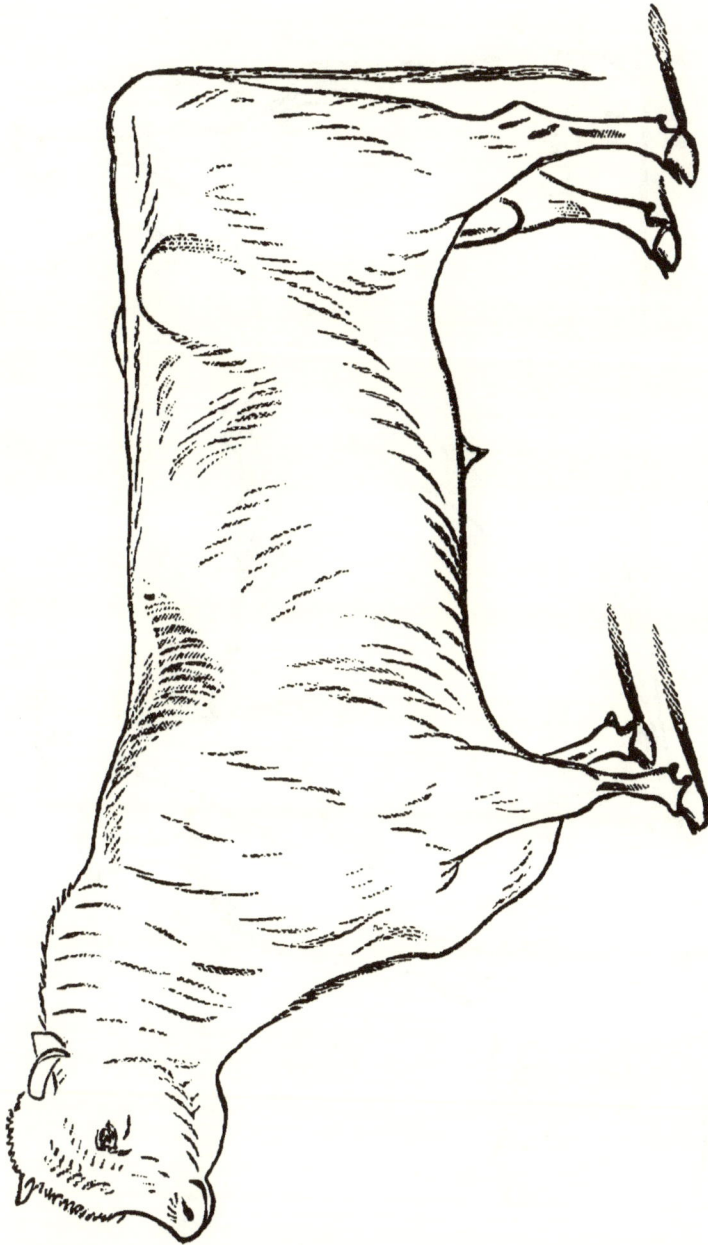

FIG. 4.—A LONG-BODIED BULL.

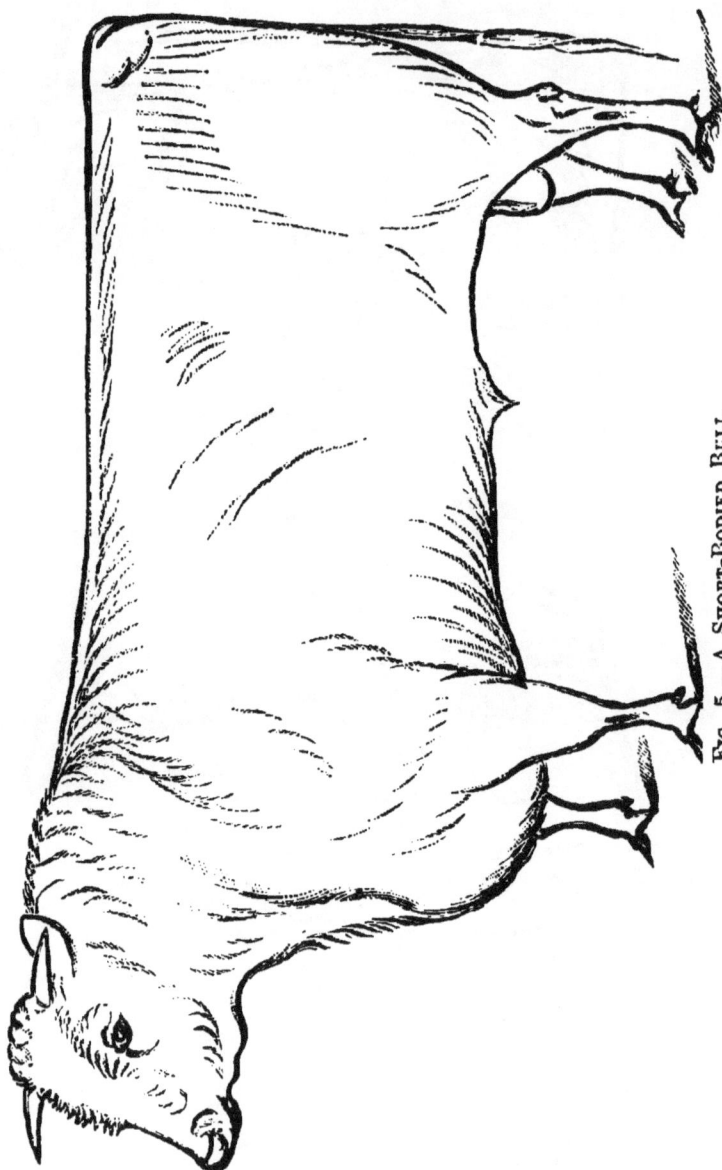

FIG. 5.—A SHORT-BODIED BULL.

well in proportion to size, but the weight, on the whole, may be deficient, and the flesh, from the excessive proportion of fat, may not be of the best quality.

Notwithstanding the objections to these very short and compact animals, as a type of the best form for the production of meat, they may be advantageously used as sires in flocks and herds that are decidedly deficient in fattening quality; the excessive tendency to the production of fat and the deficiency in muscle or lean meat being corrected by the opposite tendencies of the females with which they are coupled.

Of the two extreme types of form that have been presented the latter is to be preferred, as the defects consist only in the undue prominence or excessive development of qualities that are in themselves desirable.

When the depth and thickness of the body are in proper proportion to the length, as in Fig. 6, and the lower joints of the legs are short, so that there is not too much space between the lower line of the body and the ground, good feeding quality and early maturity may be looked for, in connection with good muscular development and flesh of the best quality, the fat being evenly distributed, while the harmony of proportions and great substance will give the greatest weight of valuable carcass.

Animals of the same dressed weight when examined on the butcher's block will be found to present great differences in the relative proportion of the cheap and the high-priced parts, and they will therefore differ greatly in actual value, without taking into account any differences that may exist in the general quality of flesh.

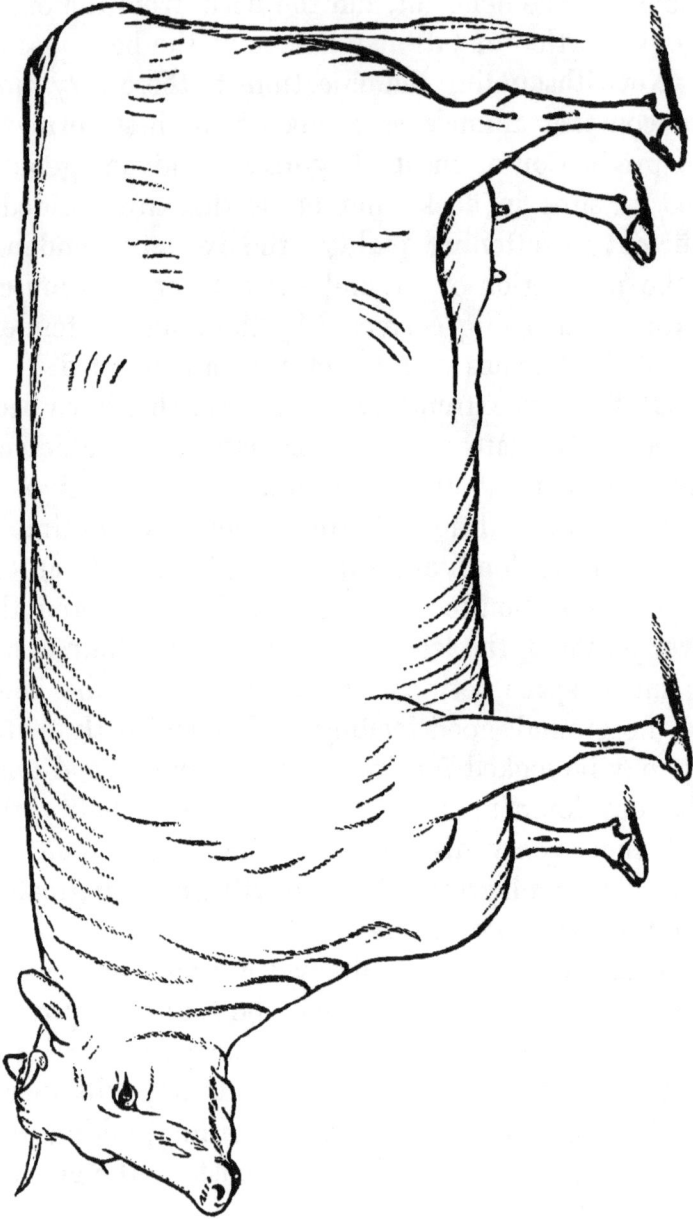

FIG. 6.—SHORT-HORN COW "BOOTH'S LANCASTER."

The cheap parts cannot, as a matter of course, be entirely dispensed with; but they should be reduced in amount as far as possible without interfering with the strength and vigor of the animal.

If a line be drawn from the shoulder-point to the knee (patella, or whirlbone), or first joint below the hip (i. e., the joint nearest the flank), in the living animal, it will be seen that the parts of the body above this line are of greater value as meat than the parts below.

In judging of the relative value of two animals that are equal in all other particulars, it will therefore be safe to say that the one giving the largest propor-tion of carcass above the line is the best for the pur-poses of the butcher.

In the best-proportioned animal it will likewise be seen that the sides of the body are filled out to the line, when applied, as above directed, in its entire length, without leaving any depressions between the line and the body at the flank and behind the arm.

This test will be found of value to the student in training the eye to detect slight variations in form, while the experienced breeder will be able to take in at a glance the conditions presented in these propor-tions without resorting to the method of actual meas-urement.[1]

[1] For observations on the general form of animals the student may profitably consult "The New Farmer's Calendar," by Lawrence, pp. 454, 455; "The Complete Grazier," p. 35; Coventry on "Agriculture," p. 174; "A Guide to Form in Cattle," by Welles; Harris on "The Pig," p. 17; *Journal of the Royal Agricultural Society*, vol. xv., p. 87; Marshall's "Midland Counties," vol. i., p. 297; Youatt on "Cattle," p. 191; "American Cattle," by Allen, p. 158; *Farmer's Magazine*, vol. xi.,

The skeleton or bony frame of the animal may next be profitably examined. As bones are of but little value, aside from the support they furnish for the soft parts of the body, they should be as small as is consistent with strength and a vigorous constitution.

Fortunately, however, the greatest strength does not depend upon size, but upon texture, the quality being of greater importance than quantity. An illustration of this may be seen in the long bones, the shaft which bears the greatest strain being small from compactness and fineness of structure, while the extremities are large and spongy, the greater surface being of use for the attachment of the tendinous terminations of the muscles.

A large, coarse bone may not only be deficient in strength, but it will increase the weight of the carcass without adding to its value.

Small bones are an indication of good feeding quality, early maturity, and superior, fine-grained flesh; while coarse, large bones, with prominent joints and angular projections of the skeleton, indicate poor feeding quality, late maturity, and coarse flesh, in connection with a large proportion of offal and cheap pieces in the carcass when reaching its final destination on the block.[1]

---

p. 98, vol. xxxvii., p. 318, vol. xxxix., p. 478, vol. xl., p. 232; Cline's "Observations on the Breeding and Form of Domestic Animals," pp. 1–8; *Quarterly Journal of Agriculture*, vol. v., p. 266, vol. vi., p. 159; Johnson's "Farmer's Encyclopædia," p. 297, and other standard works.

[1] *Farmer's Magazine*, vol. xi., p. 93, vol. xl., p. 231; *Journal of the Royal Agricultural Society*, vol. xv., p. 87; Young's "Eastern Tour," vol. i., p. 3.

" Bakewell strongly insisted on the advantage of
small bones, and the celebrated John Hunter declared
that small bones were generally attended with corpu-
lence in all the various subjects he had an opportunity
of examining." [1]

Mr. Henry Cline, an English surgeon, says : " The
strength of an animal does not depend on the size of
the bones, but on that of the muscles. Many animals
with large bones are weak, their muscles being small.
Animals that were imperfectly nourished during
growth have their bones disproportionately large. If
such deficiency of nourishment originated from a constitutional defect, which is the most frequent cause,
they remain weak during life. *Large bones*, therefore, generally *indicate an imperfection in the organs
of nutrition.*" [2]

The parts of the animal that are not deeply covered with flesh—as the head, legs, and tail, together
with the horns, when present, and the hoofs, although
of but little value in themselves—furnish the best
indications of the size, texture, and proportions, of the
bones throughout the entire system; and in the improved breeds they give an expression of refinement
and high quality to the otherwise massive structure
of the general organization.

Improvements in this direction, however, have a
limit that cannot be safely passed, as an excessive refinement of the bony tissues is often accompanied by
a delicacy of constitution that predisposes the system
to disease from exciting causes that would have little

---

[1] Sinclair's "Code of Agriculture," p. 88.
[2] " Breeding and Form of Domestic Animals," p. 7.

or no effect upon animals that have not been subjected to artificial conditions in their management.

It does not follow, however, that the constitution
is impaired by diminishing the size of the bones, or
that an increase in their size adds to its general
vigor.

All the best qualities of the improved breeds, as
has been shown elsewhere, have been obtained by
artificial treatment, that tended to disturb the equilibrium of the system, and produce changes in the
functional activity of the most important organs that
give rise to modifications of the structure, that are not
observed under what may be called the normal conditions of existence.

A greater degree of refinement and delicacy of
one set of organs involves a similar change in other
organs, through the influence of the same modifying
agencies which affect the entire system; and these,
when acting in excess, may produce a sensitiveness or
delicacy of the organization as a whole, that we recognize as a defect of constitutional vigor.

An excessive refinement of the bones would therefore indicate a delicacy and over-refinement of the
general system.

In discussing the details of external form, with
reference to the qualities indicated by peculiarities in
the development of particular parts, we will, in the
first place, examine the points of the improved Short-
Horns, as they may be fairly assumed to represent the
type of the meat-producing breeds in their most important characters.

The peculiarities of other allied breeds will only

be noticed comparatively when they differ materially from the typical form under consideration.

The "points" or parts of the animal that require attention are marked on the outline figure of a Short-Horn cow (Fig. 7): *A*, forehead; *B*, face; *C*, cheek; *D*, muzzle; *E*, neck; *F*, neck-vein; *G*, shoulder-point; *H*, arm; *I*, shank; *K*, elbow; *L*, brisket; *M*, shoulder-blade; *N*, crops; *O*, chine; *P*, loin; *Q*, hips; *R*, rump; *S*, sacrum or crupper-bone; *T*, buttock; *U*, thigh or gaskin; *V*, flank; *W*, plates; *X*, hock; *Y*, throat; *Z*, fore-flank; *qr*, quarter; *wh*, patella or whirlbone.[1]

The head should be small in proportion to the body; the frontal bone broad, without coarseness; the forehead slightly concave from the prominence of the rim of the orbits; and the face gradually tapering from the eyes to the muzzle, which should be fine, with a well-developed nostril, which indicates an ample development of the air-passages.

The jaw should be clean and free from folds of

---

[1] As the terms chine and whirlbone are often improperly used to indicate other parts of the body, it may be well to define them more particularly. Chine, in its general signification, means back; but it has long been used by breeders to indicate that part of the back between the neck and the loins. The back is therefore divided into three regions, viz., the chine, the loins, and the region of the sacrum or crupper-bone. The upper end of the femur has sometimes been called the whirl or round bone. (*See* Youatt on "The Horse," p. 262.) The terms whirlbone, turlbone, round-bone, knee-pan, knee-cap, and stifle-bone are, however, properly synonymous with patella, the bone developed in the tendon covering the knee-joint, or the articulation of the lower end of the femur with the head of the tibia, as marked in Fig. 7. (*See* Webster's, Worcester's, and Ainsworth's Dictionaries, and Wright's "Provincial Dictionary," vol. ii., p. 1019.)

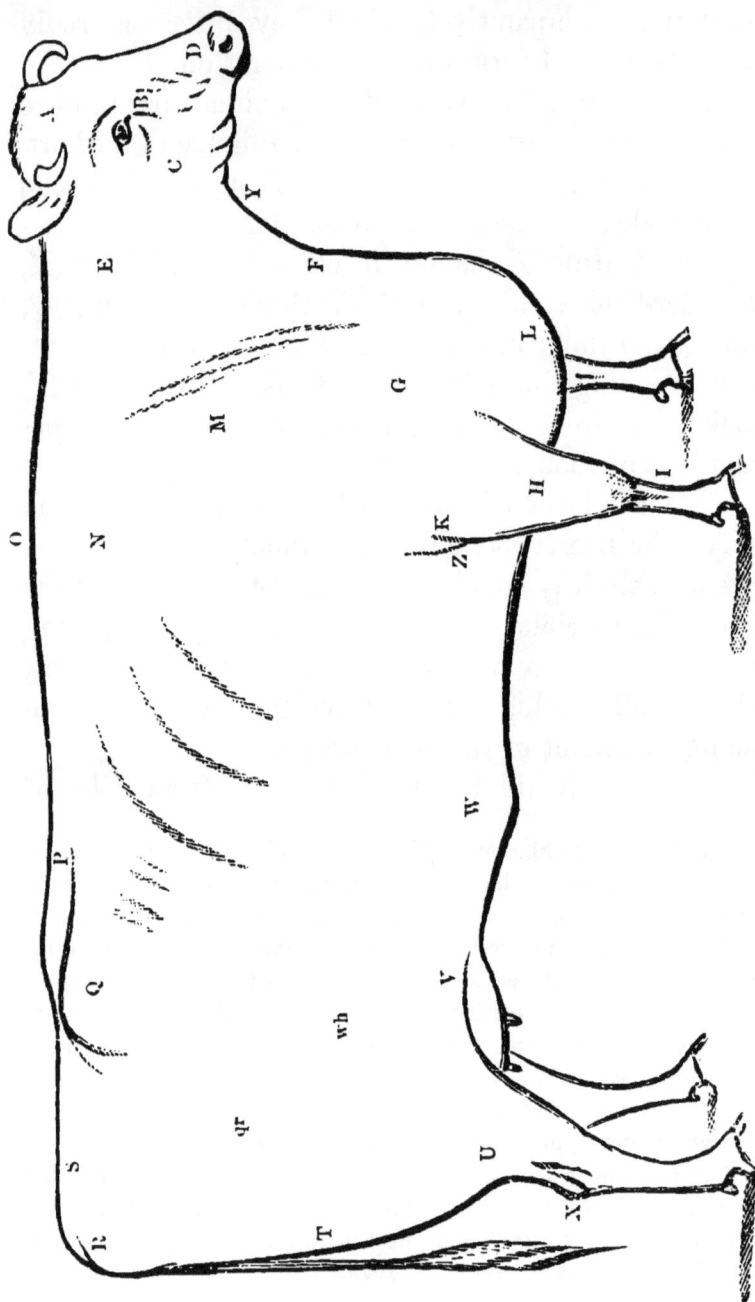

FIG. 7.—Short-Horn Cow "Pauline."

skin, and there must be no superfluous flesh on other parts of the head and face to give the animal a heavy-headed appearance. These peculiarities in the head and face are an indication of fine bones in the general skeleton, and the qualities that are usually associated with them. The eye, when prominent, bright, and clear, with a mild and gentle expression, is an indica-tion of health, with a quiet disposition and good feed-ing quality. If the eye is dull and sunken, the capil-lary circulation will be defective, and the functions of nutrition imperfectly performed; and there will not only be a deficiency in the ability to fatten, but a lack of strength and constitutional vigor. A restless and wild expression of the eye indicates a predominance of nervous action and an unquiet disposition that is not compatible with good feeding quality.

The ear should be large, without coarseness and not drooping, but with sufficient action to give a pleas-ing expression.

A drooping ear, with a general dull expression of countenance, is an indication of defective nutrition and a lack of constitutional vigor.

The horns of animals are generally supposed to be of no value, aside from their influence upon the gen-eral expression, which is considered a matter of fancy only.

Each breed has peculiarities in the size and form of the horns that are, within certain limits, character-istic; and individual breeders will choose those modi-fications of the general type that best accord with their ideas of beauty.

Notwithstanding these admissible variations in gen-

eral form, there are certain peculiarities in the devel-
ment and texture of the horns that may serve alike in
all breeds as an index of internal qualities.

The horns of the hollow-horned ruminants, including cattle and sheep, consist of a hollow bony
core developed from the frontal bone, and a sheath
or covering of true horn which, as is the case also
with the horn forming the hoofs, is composed of ag-
glutinated hairs developed from a papillary layer of
the skin.[1]

In young animals this horny sheath is thickened
by an epidermal layer that is shed as the animal grows
older, leaving the horns smoother, and at the same
time diminishing their size. In texture the horn forming the sheath of the horns, and the hoofs, seems to
be correlated with the general bony skeleton, the coat,
the skin, the flesh, and the organs of nutrition; a
clear, fine-grained texture being an indication of good
feeding quality and a general refinement of the system; while a coarse-grained, spongy texture indicates
a poor feeder and a predominance of the coarser and
less valuable parts of the carcass. Although the head
is of but little value in the slaughtered animal, its
peculiarities in the development of its appendages, as
well as its form and proportions, may, through the
correlations of structure, aid in forming an opinion in
regard to the condition of other parts of the system
that have a greater intrinsic value.

[1] "Hand-Book of Zoölogy," by Van Der Hoeven, vol. ii., p. 650;
"Cyclopædia of Anatomy and Physiology," vol. v., pp. 478–516; "Anat-
omy of the Vertebrate Animals," by Huxley, p. 327; "Comparative
Anatomy," by Wagner, p. 2.

As the neck furnishes the butcher with cheap pieces only, it might at first glance appear that it should be as light as possible, to diminish the proportion of the coarser parts of the carcass. In practice, however, it will be found true that improvements in this direction have limits that cannot be safely passed. The neck should be short, but well developed at the base, to blend symmetrically with the chine and shoulders, and thus add to the value of the fore-quarter by increasing the thickness of flesh in parts that would otherwise be defective; and it should also taper gradually toward the head, without the development of a dewlap or other indications of coarseness.

Dr. Finlay Dun says: " The distance between the ears and the angle of the jaw should be short, but the width behind the ears considerable—an important character in relation to health, as cattle with necks narrow and hollow behind the ears are defective in vigor. A well-developed neck also indicates vigor, and is especially necessary in the bull and in cattle intended for feeding. Many good milch-cows, however, have long fine necks; and, on the other hand, no cow will ever be of much value for the dairy with a short thick neck." [1]

The thickness of the neck, particularly at the base, seems also to have a direct relation to the capacity of the chest, which the feeder will consider as one of the most important parts of the animal. Many breeders of mutton-sheep prefer a thick neck, as it is usually found in connection with a capacious chest and a vigorous constitution, and thick flesh along the back. The

[1] *Journal of the Royal Agricultural Society*, vol. xv., p. 87.

upper line of the neck should be well up to the line
of the chine and loin, as a drooping neck is an indication of poor feeding quality.[1]

When the shoulders are too upright, there is often
a deficiency in the crops, and the shoulder-points are
liable to be prominent. If the shoulders are oblique
and broad at the top, they blend easily with the chine
and crops ; and, when thickly covered with flesh
throughout their entire surface, the points being obscured by the development of cellular tissue at the
base of the neck, the fore-quarter will furnish a good
proportion of valuable meat. It has been observed
that, if the shoulders are extremely oblique and narrow at the top, the upper part of the blade-bone is
not likely to be well covered with flesh.[2]

The chest contains the lungs, the heart, and the
larger blood-vessels, all of which have an important
function to perform in the process of nutrition.

The constitutional vigor, health, and feeding quality of animals, will therefore depend upon the full development of these organs, and a capacious chest that
will permit a free and vigorous performance of their
functions.

It is well known to breeders that animals with a
small chest do not fatten readily,[3] and they are remark-

---

[1] Youatt on "Sheep," p. 418 ; *Farmer's Magazine*, vol. xi., p. 98,
vol. xl., p. 232 ; *Journal of the Royal Agricultural Society*, vol. xvi., p.
36, vol. vii., p. 208.

[2] *Journal of the Royal Agricultural Society*, vol. vii., p. 208 ; *Farmer's Magazine*, vol. xxxvii., p. 319 ; Welles's "Guide to Form in
Cattle."

[3] Sir John Sebright's "Art of Improving Breeds," p. 22. W. F.
Karkeek, V. S., has assumed, on theoretical grounds, that a capacious

ably susceptible to the influence of exciting causes of disease.

The brisket, in itself considered, is of but little value, but its form is nevertheless of great importance. A narrow-pointed brisket may have a considerable development in depth, and it may be prominent when viewed from the side, but it will usually be accompanied with a chest that is too narrow at the base and lacking in depth behind the arm, a light fore-flank, and a deficiency in the development of muscle and cellular tissue between the base of the neck and the arm.

Too often this form of brisket will be found in animals with upright shoulders and defective crops.

When the brisket is broad, filling out the space on the inside of the arm in front, and its lower surface projects but little below the under-line of the body, the base of the chest will be well developed and its sides well covered with flesh, giving a good fore-flank; and there will be, as a rule, a greater compactness and uniformity in the general symmetry of the fore-quarter, and a better quality of flesh.

The hind-quarters present some peculiarities in the correlation of parts that are of particular interest to the breeder.

An extreme illustration of this may be seen in what are popularly called " pumpkin-buttocks," " lyery," or " black-fleshed " cattle (Fig. 8). The loins of these animals are very narrow, and the rump corre-

---

chest is incompatible with the rapid production of fat; but, as this assumption is based on an exploded theory of respiration, it does not require further notice. (*See Journal of the Royal Agricultural Society*, vol. v., p. 255.)

spondingly short, the tail being set on quite close to
the line of the hips. The buttocks are remarkably
full, forming a decided protuberance, that extends to

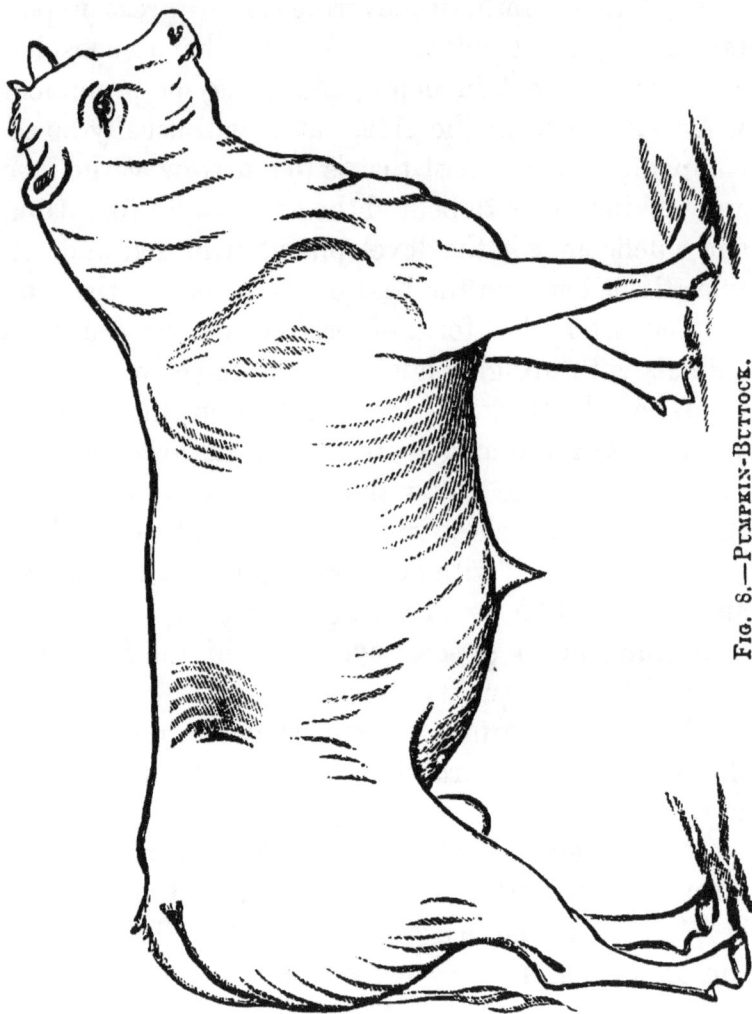

FIG. 8.—PUMPKIN-BUTTOCK.

the outer side of the thighs. With this external conformation will be found a deficiency in the formation
of fat throughout the system, the kidneys being

scarcely covered even in animals that have been highly fed, while the flesh is very dark-colored, coarse-grained, and of decidedly inferior flavor.

When any one of the above-described peculiarities of external form is present in a marked degree, the others will in all probability be found to a greater or less extent, together with an inferior quality of flesh.

In these cases it will be noticed that the best parts of the carcass are reduced to a minimum, and the coarsest parts are largely in excess.

In the best-formed animals the hind-quarters present a marked contrast to the form we have had under consideration. The loins are long and wide, diminishing the triangular space between the hips and the last ribs, and carrying the largest possible amount of choice flesh.

The hips should be broad, and the rumps long and well-filled at the sides, between the hips and the points of the rump. The tail should be set on in a line with the back, its base being broad, from a development of the transverse processes, corresponding with a similar characteristic of the loins and sacrum; while the cord, which, in its bony structure, consists of the bodies of the vertebræ only, should be fine, as an indication of small bones in the general skeleton.

The quarters from the hips and rump to the thigh should represent a vertical plane, while the twist should be full and even, without any marked protuberance of the buttocks. With this conformation will be found an abundance of fine-grained, valuable meat, while the inferior pieces will be reduced to a minimum.

The Texan steer (Fig. 9) presents a marked con-

17

trast in its form and proportions to the best type for
the production of meat. In animals like this some

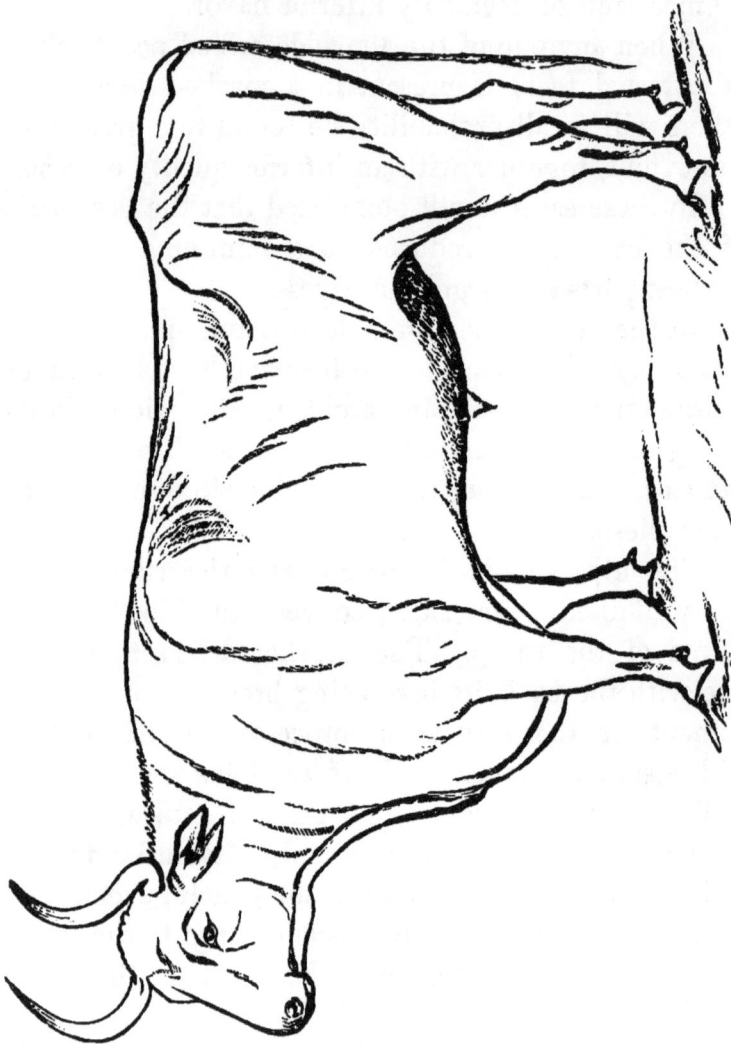

Fig. 9.—Texan Steer.

good flesh may be found in the best parts; but their
value, on the whole, is materially diminished by the
great preponderance of coarse parts in the carcass.

The "handling," "touch," or "quality," although difficult to describe, furnishes valuable indications of many of the most important characteristics of an animal.

A delicacy of the sense of touch is required to make nice discriminations by this method of examination, that can only be acquired by constant practice; and a comparison of the handling of animals that present differences in the condition of the coat and skin will need to be frequently made to prevent errors in judgment in special cases. A knowledge of the physiological principles on which this method of examination is based will be useful to the beginner, as it will enable him to appreciate those slight variations in quality that might otherwise escape his attention, and to understand more fully what he may reasonably expect to learn from its practical applications.

The activity of the capillary circulation, it is well known, is of the greatest importance in the processes of nutrition. If the materials that have been prepared by the organs of digestion for the nourishment of the system are not freely conveyed to every part of the organization, the best returns for feed consumed cannot be obtained.

As the skin is abundantly supplied with capillary blood-vessels, an examination of its properties by the "touch" will furnish the best means of ascertaining the manner in which this part of the circulatory apparatus is performing its functions.

If the capillary circulation is actively carried on in the skin, at the greatest possible distance from the large vessels of the systemic circulation, the internal parts of the organization that are more favorably situated cannot fail to be abundantly supplied with the materials required for the renovation and increase of their tissues.

What, then, are the indications of activity in the processes of circulation and assimilation in the skin that may be safely relied on as an index of the performance of these functions in other parts of the system?

The experience of practical men has enabled them to give an answer to this question that is strictly in accordance with the principles of physiological science.

Without a knowledge of the correlations of structure and function in the animal economy, that render it possible to judge of the condition of one part of the system by an examination of another, they have found by long-continued observation and experience that the quality of flesh, ability to fatten rapidly, and constitutional vigor, are uniformly reflected in the peculiarities of the coat and skin.

As it is difficult, if not impossible, to describe the slight variations in " touch " that represent marked differences in quality, so that they can be readily recognized, without practical illustrations on the living animal, we can only give a general outline of the conditions to be observed, leaving the student to gain a knowledge of details by actual experience. The first point to which attention should be directed in applying the test of "touch" is the hair, which we have already seen is correlated with the true horn of the horns and hoofs.

A fine, long, and mossy coat, that is soft under the hand, is an indication of a good feeder, and the fat, as a rule, will be well distributed, giving a good quality of fine-grained, marbled flesh. If the coat is short and fine, the animal may feed well; but there will be a tendency to the formation of internal fat, instead of that uniform distribution throughout the system that is desirable.

A harsh, coarse, wiry coat is an indication of poor feeding quality and late maturity.

In animals of good quality the skin is soft and elastic, of moderate thickness—the latter point, however, varying somewhat with the breed—yielding readily to the fingers when the animal is in moderate condition, but increasing in firmness and substance as the animal "ripens," from the ample development of fat in the cellular tissue.

A harsh, hard, and unyielding skin, in which the capillary circulation is always impaired, indicates a slow feeder and an inferior quality of flesh; while, in the opposite extreme, a thin, flabby skin, that can be readily raised in loose folds, denotes a weak constitution, and soft, oily fat, in connection with coarse, stringy flesh, that is readily recognized on the block by its lack of firmness. In the last-described condition the skin may be well supplied with capillary vessels; but the circulation is not vigorous, and it is liable to be disturbed by the slightest exciting causes. The extremes of softness and harshness represent widely-different conditions of the circulation, that are not compatible with a vigorous and efficient performance of the function of nutrition.

The principle of correlation may likewise be traced in animals that are used for work, or for the production of milk or of wool.[1] In these cases, however, the relations of particular parts to the general usefulness of the animal for its special purpose, aside from the indications of constitutional vigor that are the same in all animals, have not been as fully determined as they have in the meat-producing breeds, so that there are many details of the organization that need more extended observation and study with reference to the applications of this law of the organization.

The kind and amount of labor that can best be performed by an animal will largely depend upon the proportions of its body and limbs.

Temperament and constitutional power are of great importance in all forms of labor, as they determine the efficiency of the power applied; but they cannot act to the best advantage unless the organs of locomotion are adapted by a proper proportion of their parts to the work they are required to do.

The bones of the legs form a series of levers that are moved by appropriate muscles, which are in turn brought into activity through the influence of the nervous system. If the proportions and relative position of these levers make them act at the greatest disadvantage in the performance of a given task, the muscles that constitute the motive power, and the

---

[1] Virgil and Columella recognize the principle of correlation when they advise that a ram with a "black or spotted" tongue be rejected, as his lambs are liable to be spotted with black (Virgil, "Georgics," book iii., p. 80; Columella's "Husbandry," book vii., chap. iii., p. 306).

nervous system that brings them into action, will be subjected to a degree of tension that must impair their ability to sustain continued action, and diminish their durability. When great activity or a high rate of speed is required, the upper bones of the legs should be long and the lower bones comparatively short, together with an oblique shoulder that allows the greatest range of motion to the forearm.

For heavy-draught purposes the shoulder may be more upright, as strength rather than freedom of motion is required. A broad, flat limb, with well-developed joints, will have advantages in leverage over one that is round, from the better position of the tendons that transmit the power supplied by the muscles.

The so-called "milk-veins" of the dairy-cow are superficial blood-vessels, that represent in their development the general condition of the circulatory apparatus throughout the system, and the consequent tendency to the secretion of milk. The "escutcheon" of Guenon,[1] although perhaps not so infallible an index of milking qualities as it has been claimed to be, is undoubtedly correlated with the milk-producing function, and may therefore be of use, in connection with other points, in estimating the probable value of an animal for the dairy. From the complex relations of the various parts of the living animal, it will be seen that any single indication of quality cannot, in all cases, be assumed to represent the tendencies of the organization as a whole, for the obvious reason that the dominance of some other condition or charac-

[1] Frequently called the "milk-mirror" by other writers.

ter may obscure the relations existing between it and the organs with which it is correlated.

A single illustration will be sufficient to explain the apparent exceptions to the law of correlated structure and function that are frequently observed. A trotting horse, for example, with an extraordinary development of vital power, may be remarkable for its speed, notwithstanding a disproportion in its organs of locomotion, the defect in its structure being overcome by an excess of power. If, on the other hand, its limbs have the best possible proportions for rapid trotting, and its vital or nervous energy is deficient. it may fail in its performance, notwithstanding the perfection of its external conformation.

It does not follow from cases like these that the proportions of the limbs are a matter of indifference in the development of a high rate of speed, as better results would undoubtedly have been obtained in both cases if the defect had not been present.

CHAPTER XVII.

SELECTION.

THE intelligent breeder will make a rigorous selection of breeding-stock in accordance with a well-defined and consistent standard of excellence.

When Lord Rivers was asked how he succeeded in breeding such fine greyhounds, he replied, "I breed many, and hang many." [1]

The writer asked the late Edwin Hammond what proportion of the rams bred by himself he would be willing to use in his own flock, and he answered, promptly, "Not one in three hundred."

Mr. Dickson, in his remarks on "Selection," says, "He will prove himself the most successful breeder who can select with the most correct judgment;" [2] and it is undoubtedly true that the success of the masters of the art, who have made our improved breeds what they are, has been largely the result of the extraordinary judgment and skill with which they made their selections.

Aside from the agencies that are made use of in improving the qualities of animals, which have been pointed out in the chapter on "Variation," the art of

---

[1] *Gardener's Chronicle*, 1853, p. 45.
[2] *Quarterly Journal of Agriculture*, vol. vii., p. 248.

breeding may in fact be epitomized in the one word
" selection," which involves the application of every
established principle of practice, and a consideration
of the influence of every peculiarity of form.

The animals selected must be adapted to some
well-defined purpose in the system of management,
and to the conditions in which they are placed.

The principle that was first recognized in the
selection of stock was the adaptation of size to the
physical features of the farm, and the supply of feed.
Columella notices the difference in form and disposition of cattle and sheep arising from the conditions in
which they are placed.[1]

Fitzherbert, having in mind the same influence,
says: " And take hede where thou byeste any leane
cattel or fat, and of whom, and where it was bred.
For if thou bye out of a better grounde than thou
haste thy selfe, that cattel wyll not lyke with
the." [2]

Thomas Hale, who wrote before the marked improvement in the different breeds was made, says:
" The husbandman should be acquainted with the
several breeds, that he may suit his purchase to his
land. The larger kinds are bred where there is good
nourishment, and they require the same where they
are kept, or they will decline; the poorer and smaller
kinds, which are used to hard fare, will thrive and
fatten upon moderate land.

" The husbandman is to remember here what we
have said of trees: they never thrive if transplanted

[1] Columella, book vi., chap. i., p. 257, chap. ii., p. 304.
[2] " Boke of Husbandry " (1532), p. 46.

out of a rich into a poor soil : the same holds good
of cattle.

"The husbandman should have one of these considerations in view in stocking his land, the using
them principally for breed, for milk, or for work ;
and according as either of these is his principal aim
he is to make his purchase, one breed being fitter for
one of these uses, another for another.

"He must also consider the richness of his pastures, that he may suit the breed to that also."[1]

The experience of a Lammermuir sheep-master, as
quoted by Mr. Youatt, furnishes a good illustration
of the loss involved from lack of attention to the
principle under discussion. He says : "I occupied a
farm that had been rented by our family for nearly
half a century.

"On entering it, the Cheviot stock was the object
of our choice, and, so long as we continued in possession of this breed, everything proceeded with considerable success ; but the Dishley sheep came into fashion, and we, influenced by the general mania, cleared
our farm of the Cheviots, and procured the favorite
stock. Our coarse, lean pastures, however, were unequal to the task of supporting such heavy-bodied

[1] "A Compleat Body of Husbandry" (4 vols.), second edition, 1758,
vol. ii., p. 28.

Donaldson, in his "Agricultural Biography," says : "This work
was advertised by John Bell, of Edinburgh . . . . but no other notice
of the book can be found. . . . The 'Bibliotheca Britannica' does
not contain a book of that title among the works of that author. The
libraries of the British Museum do not possess any book of that title,
and Loudon's catalogue mentions no author of that name." The quotation in the text is made from a copy of the work in my library.

sheep, and they gradually dwindled away into less and less bulk; each generation was inferior to the preceding one; and, when the spring was severe, seldom more than two-thirds of the lambs could survive the ravages of the storm." [1]

Another striking example of the same kind is related by Mr. T. Ellman, who says : " A remarkable case in point occurred in France some years ago, when I sent some Leicester sheep to a French farmer. The ewes, sixty in number, were purchased of Mr. Golding, of Beddington; the rams, four yearlings, from Sir C. Knightly.

" The wool of these sheep was enormously heavy ; the ewes cut ten pounds each, the rams fourteen pounds each. These sheep being managed after the fashion of the Normans, the wool grew less every year, that of their progeny still lighter. In six years they clipped only three pounds of very bad wool; the fourth generation became long - legged, their bodies differing from the original stock, but resembling the native-bred Norman sheep, with which they had no relationship. After this failure a South-down ram was used, and the stock improved. Yet they soon mingled with the common flocks of the country, it being found impossible to maintain these Leicester sheep upon poor soils with bad management." [2]

With reference to size, it will, without doubt, be best to follow the advice of the author of the " Report

---

[1] " On the Breeding of Cheviot and Black-faced Sheep, by a Lammermuir Farmer," p. 66 ; quoted in Youatt on "Sheep," p. 325.
[2] *Journal of the Royal Agricultural Society*, 1865, p. 406.

on the Agriculture of Argyleshire,"[1] and make selections of animals that are rather under than over the required standard, as there will then be a reasonable prospect of improvement, and a better profit from liberal feeding.

The larger breeds, on farms that are naturally productive, have, however, in many instances proved a failure, from defects in the system of management. It cannot, with reason, be expected that the larger improved breeds will return a satisfactory profit when subjected to the same treatment that the common stock receives on the average farms of the country; and it is also quite certain that the effects of such management will be manifest in a rapid deterioration in their most valuable characteristics.

The high development of special qualities in our improved breeds, which have been obtained, as we have shown, by artificial treatment, has unavoidably diminished their hardiness, and unfitted them to withstand the effects of privation and exposure.

In the process of " natural selection " that prevails among wild species, those that are feeble or unhealthy die from exposure, and the masters of the herd attain their position by their superior strength and powers of endurance. The standard of excellence in such cases is constitutional stamina and power, and the elements of deterioration are strictly excluded.

---

[1] " Survey of Argyleshire," p. 242. *See* also on the same subject Low's " Domestic Animals," p. 264 ; Lawrence on " Cattle," p. 27 ; Coventry's " Agriculture," p. 182 ; " Complete Grazier " (sixth edition), p. 36 ; " Code of Agriculture," by Sinclair, pp. 96–100 ; " Survey of Middlesex," p. 407 ; Cline on " Breeding and Form," p. 12.

In the improvement of all domesticated varieties that are not intended for work, selections are made on a different basis, that is not favorable to the development of the greatest constitutional vigor. Take the meat-producing breeds, for example, and examine carefully the tendencies of the process of improvement. Early maturity is required, and a liberal system of feeding is practised; the wants of the animal are anticipated, and it is protected from the inclemencies of the seasons. The animal must have a quiet disposition to be a good feeder, and the treatment it receives tends to promote a habit of " masterly inactivity."

The best quality of flesh and a large proportion of choice parts are desirable, and a certain refinement of the system is the result of the efforts to obtain them.

Thus, step by step as we trace the process of improvement, we find the required conditions are unfavorable to the development or retention of constitutional vigor.

It is asserted by Prof. Tanner that, in the improved breeds, " the lungs and liver are found to be considerably reduced in size when compared with those possessed by animals having perfect liberty;" [1] and this he attributes to the lack of active exercise, which is required for the symmetrical development of the system.

The breeder must not lose sight of the tendency to undesirable variations in making his selections, or the defects of his stock may impair or even overbalance the advantages arising from their good qualities.

[1] " Transactions of the Highland Agricultural Society," 1859-'61, p. 322.

The diminution of hardiness that results from the development of the best feeding quality must not be allowed to proceed so far as to become a predisposing cause of disease.

Any inherited predisposition to disease must in like manner be carefully avoided, and the best sanitary conditions should prevail in the system of management.

The milking qualities of the meat-producing breeds have been too generally neglected, and many breeders have been led to believe that the tendency to lay on fat is directly antagonistic to the secretion of milk; and that there is an incompatibility in the active exercise of these two functions. This extreme view of the relations of the two functions is based upon certain well-ascertained facts, that do not, however, represent the whole truth.

If the attention of the breeder is directed exclusively to the development of either of these functions, the effect will be to diminish the activity of the other; and it is also well known that the peculiarities of form that indicate the best feeding quality are not the same as those obtained when the production of milk is the leading or sole object, the natural correlations of form and function in the two cases being quite different.

These facts do not, however, warrant the assumption that the two qualities, in a high degree of excellence, cannot be combined in the same animal. The possibility of such a combination of characters has been abundantly demonstrated by experience. Quite a number of animals, representing several different breeds and their grades, have come under my observation, in which good feeding qualities were associated with more than average excellence for the purposes of the dairy; and there are many similar instances on record.

Where a combination of the two qualities is the object, one of them should be made the leading or dominant character, by selections, with reference to form, in accordance with the law of correlation; while the other or secondary quality is secured through the influence of modified habits, that are ingrafted, as it were, upon the typical characteristics of the leading quality. For example, if the production of meat is the leading object, selections should be made to secure the form and proportions that experience has shown to be the best adapted to that particular purpose; while the abundant secretion of milk, which is the secondary object, may be developed as a habit of the system, notwithstanding the bias of the organization, from peculiarities of form, to the production of flesh.

A different typical form will be desirable when milk is the leading object. But with it the feeding quality may be developed to a considerable extent by an abundant supply of feed, without detracting from the value of the animal for the purposes of the dairy. In both of these typical forms, in which the com- bined qualities are developed, the energies of the sys- tem may be largely devoted to the secretion of milk during the period of lactation, and at other times to the production of flesh, so that there is an alternation in the exercise of the two functions that adds to their efficiency, from the concentration of the powers of assimilation upon a single function.

The combination of two qualities that are correlated with opposite peculiarities of form does not disprove the law of correlation, or diminish the practical value of its application. The physiological tendencies of the system, arising from correlations of form and function, are important aids to the breeder in developing a single character or quality in harmony with them; but he may nevertheless succeed in developing and retaining certain qualities, that are not strictly in harmony with the peculiarities of form, through the superior influence of modified habits and judicious selection.

Sir John Sebright recognizes this principle when he says: " It is well known that a particular formation generally indicates a disposition to get fat, in all sorts of animals; but this rule is not universal, for we sometimes see animals of the most approved forms who are *slow feeders,* and whose flesh is of a bad quality, which the graziers easily ascertain by the *touch.*" [1]

Such cases are undoubtedly rare; but their occasional occurrence is sufficient to show that the law of correlation may become latent in particular details through a preponderance of other influences.

From the plasticity of the animal organization, and its susceptibility to variation under the influence of surrounding conditions and methods of management, the breeder is enabled to obtain not only such modifications of any single characters as he may desire, but a combination of qualities which at first sight might appear to be incompatible.

The relation of the function of reproduction to the

[1] " Art of Breeding," p. 21.

secretion of milk, that has already been noticed, should not be overlooked in this connection, as an improvement in the exercise of one of these functions may have a tendency to increase the activity of the other. It is frequently more difficult to avoid defects than to secure a predominance of desirable qualities. Sir John Sebright very truly says: "We must observe the smallest tendency to imperfection in our stock the moment it appears, so as to be able to counteract it before it becomes a defect, as a rope-dancer, to preserve his equilibrium, must correct the balance before it is gone too far, and then not by such a motion as will incline it too much to the opposite side. The breeder's success will depend entirely upon the degree in which he may happen to possess this particular talent." [1]

The impaired fecundity of certain families in the improved breeds may be attributed to the neglect of this principle.

As the fecundity of animals is determined to a great extent by heredity, selections from prolific families will be found advantageous, while the opposite practice will finally result in disappointment.

Animals having the same constitutional tendencies, and kept under the same artificial conditions, may fail to breed when coupled together; but, as they prove fertile when coupled with animals of other families, the procreative function has not been lost, but made latent by conditions unfavorable to its action.

It has been suggested by Sir John Sebright [2] that

[1] "Art of Breeding," p. 6.
[2] Ibid., pp. 16, 17; "American Cattle," by R. L. Allen, p. 206.

it would be desirable to separate closely-related animals, and subject them to different conditions of food
and climate, that their development in all particulars
should not be the same. A tendency to a loss of fecundity may be corrected by this method; but the
conditions to which the animals are subjected should
not differ so widely as to destroy the characteristics
of the family that it is desirable to retain.

A defective performance of the function of reproduction may frequently be corrected by suitable selections within the family, without resorting to a change
of conditions or an infusion of other blood.

As an illustration of the manner in which the
latent function is made active, let us take the case of
two animals, kept under the same conditions, that are
closely related, so that their dominant characteristics
are essentially the same.

The male may exhibit the family defect of a tendency to impaired fecundity; while the female, with
the same general bias of the system, may be a good
milker.

When bred together, the acquired quality of se-
creting an abundant supply of milk may supplement
the conditions that give rise to the family defect, and
restore the balance of the organization, so that the
function that was comparatively latent in the parents
may become active in their offspring.

It has been claimed that Duke of Airdrie (12730)
(*see* Diagram 2) owes his superiority as a sire to characters inherited in accordance with this principle.

Duke of Gloster (11382), his sire, was not remarkably prolific, and there was a marked peculiarity in

the character of his offspring: the bulls, which were uniformly good, seemed to inherit the qualities of their grandsire, Grand Duke (10284), a superior animal; while the heifers, which were not so good, resembled their grandam. Duchess of Athol, the dam of Duke of Airdrie (12730), was a superior animal, and an excellent breeder.

Duke of Airdrie proved a good getter of both males and females, the defects of his sire having been apparently supplemented by the good qualities of his dam, although they were closely related, so that his inherited qualities were fully in equilibrium.

From the practical difficulty of making selections with reference to peculiarities that would properly supplement each other, it has been proposed to select animals that resemble each other closely in the essential or constant characters of the family, but that differ in the variable or non-essential characters: as, for instance, in the Short-Horns a difference in the color of the parents, that are alike in other particulars, may aid in restoring an impaired condition of the procreative functions by supplementing the divergent characters.

" Regard should not only be paid to the qualities apparent in animals selected for breeding, but to those which have prevailed in the race from which they are descended, as they will always show themselves, sooner or later, in the progeny; it is for this reason that we should not breed from an animal, however excellent, unless we can ascertain it to be what is called *well bred*, that is, descended from a race of ancestors who have, through several generations, possessed in a high degree the properties which it is our object to obtain." [1]

The importance of pedigree in the study of ancestral characters need only be noticed in this connection, as it has been discussed in the preceding chapters.

In order to avoid any undesirable atavic tendency, Sir John Sebright recommends as an additional precaution to " try the young males with a few females, the quality of whose produce has been already ascertained; by this means we shall know the sort of stock they get, and the description of females to which they are best adapted." [2]

As the male, from the number of his progeny, has a preponderating influence in determining the characteristics of the flock or herd, the greatest care should be exercised in his selection.

He should be more highly bred than the females with which he is coupled, to insure prepotency in the transmission of his qualities, and his merits as an individual should add to the reputation of the long line of ancestry from which he is descended.

Breeders of pure-bred stock are aware of the importance of securing males of extraordinary excellence in every respect, and high prices are accordingly paid for the best representatives of favorite families.

Those who use males of their own breeding select

[1] "Art of Breeding," p. 7.

[2] *Loc. cit.*, p. 7. (*See* also Sinclair's " Code of Agriculture," p. 98.)
According to Arthur Young, it was the practice of the late Duke of Bedford to place " every ram with the lambs got by him the preceding year, in distinct pens, that he might not only examine the ram himself, but also his progeny, before he determined what ewes to draw off for him " ("Farmer's Calendar," p. 568).

them from a family that is more highly bred than the rest of their stock.

It seems to be the prevailing opinion that almost any pure-bred male will answer the purpose of those who are breeding grades, and comparatively few think of making their selection in accordance with any definite system.

In the improvement of grade-stock the breeder should have clearly-defined ideas of the kind of animal he would produce, and the rules of the art that have been established by the breeders of pure-bred animals will be found the safest guides in his practice.

High-bred males, of the particular type it is proposed to establish, will impress their own characteristics upon their offspring with greater certainty and uniformity than those that, although of pure blood, have been bred from an admixture of a variety of elements without reference to any definite standard.

Even for the purpose of improving grades it will be found more profitable to select a high-bred animal of superior merit than to use one that cost half the money, whose qualities are not so well defined.

As the dangers of in-and-in breeding are not so great in breeding grades (a pure-bred sire always being used) as in breeding pure-bred stock, a well-bred male that is free from defects may be used upon his own get with advantage, while a similar practice with an inferior animal would not be desirable. In some of the best grade-herds that have come under my observation, in-and-in breeding (on the part of the sire) has been practised for several generations without any indications of unfavorable results. The sires in these cases have been animals of strong constitution, and apparently free from inherited predisposition to disease.

In the improvement of grades, as well as pure-bred animals, the selection of breeding-stock must go hand-in-hand with a judicious system of feeding and management, as the artificial characters which are impressed by the male upon his offspring can only be retained through the influence of essentially the same conditions that originally produced them.

CHAPTER XVIII.

THE duration of the period of gestation in mammals is apparently determined by various causes that we are as yet, from the obscurity of their action, unable to define.

That it bears some relation to the size of the animal is shown by the following instances, which have been compiled from various sources ;[1] and Mr. Darwin states that it has been observed in Germany that " the period of gestation is longer in large-sized than in small-sized breeds of cattle."[2]

The period of gestation is approximately as follows : Elephant, twenty to twenty-three months ; giraffe, fourteen months ; dromedary, twelve months ; buffalo, different varieties, from ten to twelve months ; ass, twelve months ; mare, eleven months ; cow, two hundred and eighty-five days ; bear, six months ; rein-

---

[1] Van Der Hoeven's " Zoölogy," vol. ii. ; " Æconomische neugikund Verhandl," quoted in Johnson's " Farmer's Cyclopædia," and *Quarterly Journal of Agriculture,* vol. x., p. 287; Dunglison's " Human Physiology ;" Johnson's " Cyclopædia," article " Gestation," by Dr. E. R. Peaslee ; " Encyclopædia Britannica," article " Animal Kingdom," by Prof. Wilson.

[2] " Animals and Plants under Domestication," vol. ii., p. 387.

deer, eight months; monkeys, seven months; sheep and goat, five months; sow, four months; beaver, four months; lion, one hundred and eight days; puma, seventy-nine days; dog, fox, and wolf, sixty-two to sixty-three days; cat, fifty days; rabbit, thirty days; squirrel and rat, twenty-eight days; Guinea-pig, twenty-one days.

A similar relation may be traced in the period of incubation in birds, which is as follows: Turkey, twenty-six to thirty days; Guinea-hen, twenty-five to twenty-six days; pea-hen, twenty-eight to thirty days; ducks, twenty-five to thirty-two days; geese, twenty-seven to thirty-three days; hens, nineteen to twenty-four days, or an average of twenty-one; pigeons, sixteen to twenty days; canary-birds, thirteen to fourteen days. Mr. Wright remarks that " cold weather, or a prevailing east wind, will lengthen the time a day or more, while warm weather and an attentive sitter will hasten it; stale eggs also hatch later than fresh."[1]

He also states that the small breeds require less time than the large breeds; " Hamburgs generally hatch at the expiration of the twentieth day, and Game Bantams often even on the nineteenth." Mandarin and Wood ducks " usually hatch in about twenty-five days; but something depends upon whether the eggs are set under hens which, owing to the greater heat of their bodies (at least we suppose so, reasoning generally), hatch from one to two days earlier than if the same eggs are set under their natural parent."[2]

[1] " Book of Poultry," p. 49.   [2] *Loc. cit.*, p. 556.

18

The period of incubation is said to be shortened when hens' and ducks' eggs are set under a turkey.[1]

The following statistics will indicate the variations that are liable to occur in the period of gestation of different animals. M. Tessier, who continued his observations for forty years,[2] has made the most valuable collection of facts in relation to this subject, which will be quoted under their appropriate heads.

According to Youatt, the average period of gestation in the mare is eleven months, but it may be diminished five weeks or extended six weeks.[3]

Of 582 mares reported by M. Tessier, the shortest period was 287 days, the longest 419, and the average 330 days.[4]

M. Gayot has recorded the period of gestation for twenty-five mares, the shortest period being 324 days, the longest 367 days, and a mean of 343 days.[5]

It will be noticed that the range of variation is less in the cases observed by M. Gayot, while the average period exceeds that of M. Tessier's observation thirteen days. Of 575 cows observed by M. Tessier—

21 calved between the 240th and 270th days, the mean time being 259 days.

544 calved between the 270th and 299th days, the mean time being 282 days.

---

[1] "Farmer's Cyclopædia," p. 562; *Quarterly Journal of Agriculture*, vol. x., p. 287.

[2] Carpenter's "Human Physiology," p. 983.

[3] "The Horse," p. 222.

[4] Johnson's "Farmer's Cyclopædia," p. 562; "Encyclopédic pratique de l'Agriculteur," vol. viii., p. 298.

[5] Ibid.

10 calved between the 299th and 321st days, the mean time being 303 days.[1]

The extremes here given were not changed when the number observed was extended to 1,131 animals, but the results as to the average are not stated.[2]

Earl Spencer has recorded the period of gestation in 764 cows with the following result: Least period, 220 days; mean, 285 days; longest period, 313 days. But he remarks that he has "not been able to rear any calf produced at an earlier period than 242 days."[3]

As the table published by Earl Spencer is of interest in many particulars, it is copied in full:

Number of Days of Gestation.	Cows.	Cow-Calves.	Bull-Calves.	Twin Cow-Calves.	Twin Bull-Calves.	Twin Cow and Bull Calves.
220.............	1	..	1	..	..	..
226.............	1	1	..	..	..	..
233.............	1	..	1	..	..	..
234.............	1	..	1	..	..	..
235.............	1	1	..	..	..	..
239.............	1	1	..	..	..	..
242.............	1	..	1	..	..	..
245...........	2	2	..	..	..	..
246.............	2	..	2	..	..	..
248.............	1	1	..	..	..	..
250.............	1	1	..	..	..	..
252.............	2	..	2	..	..	..
253.............	1	..	1	..	..	..
254.............	1	1	..	..	..	..
255.............	2	..	2	..	..	..
257.............	2	1	1	..	..	..
258.............	8	1	2	..	..	..
259.............	1	..	1	..	..	..
262.............	1	..	1	..	..	..
263.............	2	..	2	..	..	..
266.............	1	..	..	..	1	..
268.............	2	2	..	..	..	..
269.............	2	..	1	..	..	1
270.............	5	2	1	1	..	1
271.............	6	5	1	..	..	..
272.............	8	1	1	..	1	..
273.............	8	2	1	..	..	..

[1] "British Husbandry," vol. ii., p. 438; "The Complete Grazier," p. 47.

[2] Youatt on "Cattle," p. 527.

[3] *Journal of the Royal Agricultural Society*, vol. i., pp. 166, 167.

Number of Days of Gestation.	Cows.	Cow-Calves.	Bull-Calves.	Twin Cow-Calves.	Twin Bull-Calves.	Twin Cow and Bull Calves.
274	5	..	5	..	..	..
275	5	2	2	..	1	..
276	15	7	6	..	1	1
277	14	10	2	1	..	1
278	13	11	4	1	..	2
279	82	16	11	3	..	2
2-0	85	15	20	..	..	..
2-1	89	20	18	..	..	1
2-2	47	26	20	1	..	..
2-3	54	30	24	..	..	..
2-4	66	33	33	..	..	..
2-5	74	29	43	..	..	2
286	60	22	38	..	..	..
287	52	25	27	..	..	..
288	43	13	28	..	1	..
239	45	20	25	..	..	..
290	23	10	13	..	..	..
291	31	9	22	..	..	..
292	16	5	11	..	..	..
293	10	1	9	..	..	..
294	8	1	7	..	..	..
295	7	8	4	..	..	..
296	6	2	4	..	..	..
297	2	1	1	..	..	..
299	1	..	1	..	..	..
304	1	1	..	..	..	..
305	1	1	..	..	..	..
306	3	8	..	..	..	..
307	1	1	..	..	..	..
313	1	1	..	..	..	..

L. F. Allen reports the period of gestation in fifty cows for a single year as follows: Shortest period, 268 days; mean, 284 days; and longest period, 291 days.[1]

These cows were "Short-Horns, Herefords, Devons, and their grades, and common ones;" but no difference was noticed that could be attributed to the breed.

C. N. Bement[2] reports the period of gestation for five years in his herd of cows, consisting of "Durham, Devon, Hereford, Ayrshires, and grades," which will be found in the following table:

[1] "American Cattle," p. 253.
[2] *The Cultivator*, 1845, p. 207.

YEAR.	No. of Cows.	COW CALVES.				BULL CALVES.			
		No.	Shortest Period.	Longest Period.	Average Period.	No.	Shortest Period.	Longest Period.	Average Period.
1839.......	14	3	....	....	284	11	....	....	280
1840.......	13	6	213	336	273	7	278	289	299
1841.......	11	8	277	292	286	3	284	299	283
1842.......	13	4	280	286	284	9	281	294	287
1843.......	11	5	276	286	282	6	277	290	282
Total....	62	26	....	....	283	36	....	....	283

Mr. Bement had "doubts as to the correctness" of the shortest period given—213 days—which is evidently an error, as he has in no other instance observed a period "below 260 days."

The minimum of Mr. Bement's observations, if not an error, must be considered decidedly premature.

In M. Tessier's observations on sheep,[1] of 912 ewes—

140 lambed between the 146th and the 150th days; mean time, 148 days.

676 lambed between the 150th and the 154th days; mean time, 152 days.

96 lambed between the 154th and the 161st days; mean time, 157 days.

In 420 ewes under the observation of M. Magne, at Alfort, the period of gestation was—

149 days for   80
148 " "   68
150 " "   55
147 " "   55
151 " "   49
146 " "   30
152 " "   23

[1] "The Complete Grazier," p. 238; "British Husbandry," vol. ii., p. 457; Youatt on "Sheep," p. 496.

145 days for	.	.	.	.	.	22
144 " "	.	.	.	.	.	15
153 " "	.	.	.	.	.	13
154 " "	.	.	.	.	.	7
156 " "	.	.	.	.	.	3
143 " "	.	.	.	.	.	2

The extremes being 143 and 156 days, and the entire period in three-fifths of the flock from 147 to 150 days.[1]

In 1814 and 1815 M. Morel de Vinde recorded the period of gestation in 462 ewes as follows :

153 days in	.	.	.	.	.	118
152 " "	.	.	.	.	.	97
151 " "	.	.	.	.	.	81
150 " "	.	.	.	.	.	50
154 " "	.	.	.	.	.	42
149 " "	.	.	.	.	.	31
155 " "	.	.	.	.	.	18
148 " "	.	.	.	.	.	7
156 " "	.	.	.	.	.	6
157 " "	.	.	.	.	.	5
147 " "	.	.	.	.	.	4
146 " "	.	.	.	.	.	3

In more than three-fifths of these ewes the duration of gestation was from 151 to 153 days, or from three to four days longer than in the cases observed by M. Magne, and the extreme periods—146 and 157 days—are likewise more prolonged.

These variations in the length of the period in the two flocks may be attributed to a difference in breed, or in the system of management, or possibly to local influences.[2]

[1] " Encyclopédie pratique de l'Agriculteur," tome x., p. 483.
[2] Ibid.

Mr. Darwin states,[1] on the authority of Nathusius, "that merino and Southdown sheep, when both have long been kept under exactly the same conditions, differ in the average period of gestation, as seen in the following table :

Merinos	150.3 days.
Southdowns	144.2 "
Half-bred merino and Southdown	146.3 "
Three-fourths blood of Southdowns	145.5 "
Seven-eighths " " "	144.2 "

The average period of gestation in swine is about sixteen weeks. The extremes observed by M. Tessier in twenty-five sows were 109 and 123 days ;[2] and Mr. Fox has reported " ten carefully-recorded cases with well-bred pigs in which the period varied from 101 to 116 days.[3]

M. Tessier " observes that the extent of gestation is in many species extremely various, and that its prolongation does not seem to depend upon the age or constitution of the female, or upon the diet, breed, or season, or, in short, upon any known cause." [4]

It is, however, at least probable that the period of gestation is shorter in the breeds that mature early, and this may be the explanation of the difference

[1] "Animals and Plants under Domestication," vol. i., p. 123.

[2] Darwin, "Animals and Plants under Domestication," vol. i., p. 95 ; Youatt on " The Hog," p. 154.

In " The Complete Grazier," p. 299, and in "British Husbandry," vol. ii., p. 511, the extremes observed by M. Tessier are stated at 109 and 143 days ; but this is evidently an error.

[3] "Animals and Plants under Domestication," vol. i., p. 95.

[4] "British Husbandry," vol. ii., p. 511 ; "The Complete Grazier," p. 299.

observed in merino and Southdown sheep that has
been noticed.

In swine, "according to Nathusius, the period is
shortest in the races which come early to maturity;
but in these latter the course of development does not
appear to be actually shortened, for the young animal
is born, judging from the state of the skull, less fully
developed or in a more embryonic condition than in
the case of common swine, which arrive at maturity
at a later age.[1]

It seems to be the general opinion that the period
of gestation is longer with male than with female offspring;[2] but there appears to be no sufficient evidence
on record to warrant such a conclusion.

In the observations of Mr. Bement, from "those
cows that exceeded 286 days, the number of females
was seven, while that of the males was twelve. The
number of female calves produced under 283 days
was twenty-four, while that of the males was thirty-
one."[3] There was thus a larger proportion of males
in the periods above and below what may be considered an average, and it is worthy of remark that the
produce in the longest-observed period was a heifer.

The average period was the same for males and
females in 1845, was longer for females in 1839, and
longer for males in the three remaining years. The
average for the five years was 288 days for males and
283 days for females.

[1] Darwin, "Animals and Plants under Domestication," vol. i., p. 96.
[2] "British Husbandry," vol. ii., p. 438; Randall's "Practical Shepherd," p. 207.
[3] *The Cultivator*, 1845, p. 207.

Earl Spencer was inclined to believe that his observations show that "there is some foundation for this opinion. . . .

"In order fairly to try this," he says, "the cows calved before the 260th day and those who calved after the 300th ought to be omitted as being anomalous cases, as well as the cases in which twins were produced; and it will then appear that, from the cows whose period of gestation did not exceed 286 days, the number of cow-calves produced was 233 and the number of bull-calves 234; while, from those whose period exceeded 286 days, the number of cow-calves was only 90, while the number of bull-calves was 152." [1]

He neglects, however, to notice that in the entire number of births, omitting the twins, there were but 340 cow-calves to 401 bull-calves, or a large preponderance of males; and that all the calves born after the 300th day were females, while of those born before the 260th day ten were cow-calves and fifteen bull-calves.

M. Magne, on the contrary, found the period of gestation longer with ewe-lambs than with ram-lambs, and this he attributes to the greater development of the males previous to birth. [2]

The duration of gestation seems to depend also to some extent upon heredity. "It was ascertained by the late Earl Spencer that of seventy-five cows in calf by a particular bull, the average period was 288½ days instead of 280, none of them having gone less than

[1] *Journal of the Royal Agricultural Society*, vol. i., p. 168.
[2] "Encyclopédie pratique de l'Agriculteur," tome x., p. 485.

281 days, and two-fifths of them having exceeded 289 days." [1]

From the facts that have been presented it appears that the size, early maturity, and inherited tendencies, may all have an influence in determining the duration of gestation.

The wide range of variations that occur in the same family, and even in individuals, seems to indicate that there are other and perhaps more efficient influences that have escaped our attention.

[1] Carpenter's "Human Physiology," p. 982.

APPENDIX.

THE following examples of the form of record in the different herd-books are given for the benefit of persons not familiar with pedigrees.

Short-Horn pedigrees :

"(14837)   LORD OF THE VALLEY.

Red, calved August 30, 1856, bred by Mr. R. Booth, Warlaby ; got by Crown Prince (10087), dam (Red Rose) by Harbinger (10297), g. d. (Medora) by Buckingham (3239), gr. g. d. (Monica) by Raspberry (4875), — (White Strawberry) by Rockingham (2551), — by Young Alexander (2977), — by Pilot (496), — by the Lame Bull (359), — by Easby (232), — by Suwarrow (636)."—("English Short-Horn Herd-Book," vol. xii., p. 137.)

"9798   DUKE OF AIRDRIE. (12730)

[*The original progenitor of the American Dukes of Airdrie, called in Kentucky ‘ The Old Duke.’*]

Red and white, bred by R. A. Alexander, Airdrie, Scotland, and imported to his farm in Woodford Co., Ky., calved Aug. 4, 1854, got by imp. Duke of Gloster, 2763

(11382), out of Duchess of Athol, by 2d Duke of Oxford (9046), — Duchess 54th, by 2d Cleveland Lad
(3408), — Duchess 49th, by Short Tail (2621), —
Duchess 30th, by 2d Hubback (1423), — Duchess 20th,
by 2d Earl (1511), — Duchess 8th, by Marske (418), —
Duchess 2d, by Ketton 1st (709), — Duchess 1st, by
Comet (155), — by Favorite (252), — by Daisy Bull
(186), — by Favorite (252), — by Hubback (319), —
the Stanwick cow, by J. Brown's Red Bull (97)."

Allen's "American Short-Horn Herd-Book," vol. x.,
p. 107.

Numbers in parentheses refer to "English Herd-
Book," open numbers to the "American Herd-Book."

"171   DUKE OF AIRDRIE.   (12730)

Red and white, calved August 4, 1854; bred by, and
the property of, Mr. R. A. Alexander, Airdrie House,
Airdrie; got by Duke of Gloster 175, d. (Duchess of
Athol) by 2d Duke of Oxford 180, g. d. (Duchess 54)
by 2d Cleveland Lad 123, — (Duchess 49th) by Short
Tail 498, — (Duchess 30th) by 2d Hubback 281, —
(Duchess 20th) by the 2d Earl 183, — (Duchess 8th)
by Marske 358, — (Duchess 2d) by Ketton 1st, 305, —
(Duchess 1st) by Comet 128, — by Favourite 204, —
by Daisy Bull 151, — by Favourite 204, — by Hubback
280, — by Mr. James Brown's Red Bull 80."—(Alexander's "Short-Horn Record," vol. i., p. 27.)

By comparing these pedigrees with the diagrams on
pages 147 and 142 it will be seen that they are given in
an abbreviated form in the herd-books, and that each
animal appearing in the record must be separately traced
to obtain a complete list of the ancestors.

It will also be observed that the original numbers of

the bulls are used in the "American Herd-Book," while new numbers are assigned them in Alexander's "Short-Horn Record."

The dash (—) is used in each of the forms given to indicate the "next dam," and thus save space in the record.

Hereford pedigrees :

"376 COTMORE W. F., calved 1836, bred by the late Mr. T. Jeffries, by Old Sovereign (404), dam by Lottery (410). At Mrs. Jeffries's sale 1844 Cotmore was bought in for £100 ; he won, at different times, the prizes for two-year-old, three-year-old, and aged bulls at Hereford ; and the first prize for Hereford bulls at the meeting of the Royal Agricultural Society at Oxford ; Cotmore's dam, at the Grove sale 1844, was sold for £33."—(" The Herd-Book of Hereford Cattle," vol. i., p. 52. *See* page 164 for extended pedigree.)

"(3434)   SIR CHARLES.

Red with white face, calved February 14, 1867 ; bred by and the property of Mr. F. W. Stone, Moreton Lodge, Guelph, Canada ; got by Guelph (2023), dam (Graceful) by Severn (1382), g. d. (Lady) by Albert Edward (859), g. g. d. (Zephyr) by Walford (871), — (Friday the Second) by Wonder (420) — (Friday) by Commerce (354), — (Pretty Maid) by The Sheriff (356), — (Sovereign) (404)."—(" Herd-Book of Hereford Cattle," vol. vii., p. 125.)

The cows in all the above cases are identified by the name of their sire following their own. W. F. after Cotmore means white face ; in the first volumes of the

"Hereford Herd-Book" this abbreviation was used, as also M. F. for mottled face, G. for gray, etc.

Devon form of pedigree :

"Prince of Wales (105) referred to as Quartly's Prince of Wales ; calved in 1843, bred by James Quartly, the property of Earl Leicester. He won the 1st prize as best young bull in 1844, and 1st prize as best old bull in 1845, at Exeter ; and 1st prize in class 2, at the R. A. M. at Shrewsbury. Sire, Prince Albert (102) : grandsire, Hundred Guinea (56) : dam Duchess (146) by Hundred Guinea (56) : grandam Lilly, by a son of Forester, (46) out of Long-Horned Curly, bred by Mr. F. Quartly."—(Davy's "Devon Herd-Book," vol. i., p. 26. *See* page 149 for the same pedigree in tabular form.)

"466 Eveleen 5th.

Calved March 14, 1862 ; bred by the late Edward G. Faile, West Farms, N. Y. ; the property of Michigan State Agricultural College, Lansing, Mich.

Sire *Cayuga* (602) (587 E) ; 2d sire, Tecumseh (567) (535 E) ; 3d sire, Frank Quartly (205), imported ; 4th sire, Earl of Exeter (38) ; 5th sire, Baronet (6).

Dam, imported *Eveleen* (691), bred by Mr. George Turner, of Barton, England, by Earl of Exeter (38) ; 2d Dam, Ruby (1035), by Favorite (43) ; 3d Dam, Pink (952), by a son of Pretty Maid (366), and Watson (129) ; 4th Dam, bred by Mr. John Halse."—("American Devon Herd-Book," vol. ii., p. 105.)

In Devon pedigrees "The figures in parentheses with the letter E, thus (00 E) refer to Davy's third volume of 'English Devon Herd-Book.' The figures in parentheses, thus (00), refer to Davy's first and second volumes, and Howard's third volume ;" while in references to the "American Devon Herd-Book" the figures are not inclosed in parentheses.

This complication in the numbers designating recorded animals arises from the simultaneous publication, in England and America, of a third volume of pedigrees, the numbers in each being a continuation of the numbers in the first and second volumes of the "English Herd-Book." There are therefore two so-called third volumes of the "Devon Herd-Book," one English, a continuation of Davy's original series, and the other American, known as Howard's third volume.

After the publication of the latter an "Association of Breeders" started an American "Devon Herd-Book," in which American pedigrees are now recorded.

Ayrshire form of record :

"668 NETTIE.

Light red with a little white ; calved May 13, 1863 ; bred by Henry H. Peters, Southboro, Mass. ; owned by Prof. Manly Miles, Lansing, Mich.

Sire, Eglinton, 21.   Dam, Ruth, 193."

(American "Ayrshire Herd-Book," vol. ii., p. 102.)

The sire and dam only are given here, and reference to the record under their numbers is necessary to extend the pedigree.

There are now three "Ayrshire Herd-Books" published in America ; but we need not give examples of pedigrees from all of them, as the system of recording is essentially the same, the cows as well as the bulls having a distinguishing number.

The pedigrees in the record of the "American Jersey Cattle Club" are published in tabular form ; the sire

and dam, each with a distinguishing number, are alone
given.   The headings of the different columns, in which
the record is made, are as follows :

*No.—Name—Color and distinguishing marks—By
whom bred or imported— When dropped or imported—
From what place, in what vessel—Present or last owner
—Sire—Dam.*

INDEX.

THE END.

www.ingramcontent.com/pod-product-compliance
Lightning Source LLC
Chambersburg PA
CBHW032303280326
41932CB00009B/677

9 783337 143596